A New Kind of Knowledge

A New Kind of Knowledge

Evocations, Exhibitions, Extensions, and Excavations

edited by Jack Petranker

Dharma Publishing

PERSPECTIVES IN TIME SPACE AND KNOWLEDGE

Dimensions of Thought I and II
Mastery of Mind
Visions of Knowledge
Light of Knowledge
A New Way of Being
A New Kind of Knowledge

Library of Congress Control Number 2004090158

ISBN 0-89800-357-1

Cover design by Emily Jan. Labyrinth design based on the labyrinth at Chartes Cathedral.

Typeset in Adobe Trump Mediaeval and Helvetica Light. Printed and bound by Dharma Press, Berkeley, CA.

9 8 7 6 5 4 3 2 1

Dedicated to Knowledge:
open as space,
alive as time,
free from all boundaries,
always available.

Contents

EXHIBITIONS

EXTENSIONS

EXCAVATIONS

Preface

They said, "You have a blue guitar,
You do not play things as they are."
The man replied, "Things as they are
Are changed upon the blue guitar."
—Wallace Stevens

The TSK vision first emerged into the world in 1977, with the publication of *Time, Space, and Knowledge: A New Vision of Reality*. That work was the occasion for great excitement. It seemed to its first readers that something new had been brought into the world.

Twenty-five years later, the promise of TSK is as rich as ever. The vision has proved challenging; more challenging, perhaps, than those first entranced readers might have imagined. For every student of the vision who will affirm that working with the TSK books and exercises has been a life-transforming experience, there are dozens more who have never found a way in to the vision, who come away with questions that seem to them to lead nowhere and a sense that whatever is going on in the TSK books is in some fundamental sense beyond them. In one way this is absolutely right: Time, Space, and Knowledge in the sense that they are evoked in the TSK vision are 'beyond' what you or I or anyone can know. The vision invokes a knowledge that the individual knower cannot lay claim to, and this is precisely the source of its special power. Perhaps this book and its companion volume, *A New Way of Being* (being published at the same time), can help show how this 'beyondness' is consistent with the immediacy of insight and inquiry. If so, they will be serving an important function, for it is when we bring these two aspects of the vision—immediacy and 'beyondness'—together that we can see 'where to go from here'.

In *Mind over Matter: Reflections on Buddhism in the West* (p. 84), Tarthang Tulku starts a brief discussion of TSK and the role that it can play in today's world by writing:

> [T]he books in the Time, Space, and Knowledge Series . . . promote a style of inquiry that is well-suited to Dharma study and secular concerns alike. For certain individuals, including intellectuals and scientists highly skeptical toward all religions, these works, which make no dogmatic claims and establish no belief systems, may serve as a point of entry to a knowledge that goes beyond the ordinary mind.

This description may seem to confirm that the TSK books are not for everyone, and speaking in terms of the reaction they have provoked, this is certainly the case. Yet the essays and articles collected in this volume suggest that TSK has a broader reach. People with no particular training have found that the TSK vision opens up new possibilities that make their lives more fruitful, their experience more fulfilling, and the range of possibilities open to them more vast.

Because we are so committed to a way of knowing in which the self is the one that knows, TSK can seem a puzzle, since it maintains so strongly that *knowledge is not what the knower knows.* Some will find this statement puzzling, perhaps even irritating, but others will see it as an invitation. For there can be little doubt that the self-centered way of knowing at which we are all reasonably skilled does not give us the knowledge we need. It is too dedicated to positions and conditions, to identity and insistence.

Others have addressed this fundamental difficulty, calling for a new kind of knowing and a different understanding of truth, one not bound to the structures and limits of what *Love of Knowledge.* calls "technological knowledge." But TSK differs in two remarkable ways. First, it makes no attempt to challenge the self as such. Instead, it challenges time and space, and thus the structures on which the self unthinkingly depends. Second, the TSK vision does not seek to tell or persuade, but to show. And it shows not by pointing out what is already there to be seen, but by inviting us to look for ourselves. *What* we see is not so important. What matters is that we learn to see, and learn not to close down our seeing prematurely.

A preface is normally written in a single voice. In keeping, with the thought, however, that knowledge is not something to be owned, I will end these remarks with some thoughts by Ronald Purser, a contributor to this volume, that seem to me to fit especially well here, and to speak on behalf of all the authors who are represented in these pages:

> The Time, Space and Knowledge vision is truly a remarkable gift to the West. In the spirit of this volume commemorating the twenty-fifth anniversary of TSK, I would like to express my heartfelt gratitude and deepest thanks to Tarthang Tulku and his devoted staff at Dharma Publishing for making the TSK vision available.
>
> In the preface to the first TSK book, Tarthang Tulku remarked that the TSK vision in essence "wrote itself." But we all know the tremendous amount of effort and time that he has devoted to producing the TSK series of works, beginning with the foundational volume. As most of us know, Tarthang Tulku is not someone who wastes his time, nor the time of others. Thus, for him to spend so much of his time on TSK over the years is a commitment of great significance. Given the personal and professional benefits that I have received from TSK, I feel it is my responsibility to generate more knowledge that might help this visionary work be more accessible and useful to others.

Abbreviations

The following works in the *Time, Space, and Knowledge* series or the *Perspectives on Time, Space, and Knowledge* series, listed in order of publication date, have been abbreviated as indicated. All are published by Dharma Publishing. Note that the abbreviation "TSK" is sometimes also used in this volume to refer to the Time, Space, and Knowledge vision. It should be clear from context which usage is intended.

Works by Tarthang Tulku

TSK	*Time, Space, and Knowledge: A New Vision of Reality*. 1977.
LOK	*Love of Knowledge.* 1985.
KTS	*Knowledge of Time and Space.* 1990.
VOK	*Visions of Knowledge: Liberation of the Modern Mind*. 1993.

DTS	*Dynamics of Time and Space: Transcending Limits on Knowledge*. 1994.
SDTS	*Sacred Dimensions of Time and Space*. 1997

Other Works

DOT	*Dimensions of Thought: Explorations in Time, Space, and Knowledge* (2 volumes). Eds. Ralph Moon and Steve Randall. 1980.
MOM	*Mastery of Mind*. 1993.

EVOCATIONS

Any student of the TSK books will have been struck by their boundless creativity and inventiveness. Every page, almost every paragraph, contains something new, something that takes what has been presented previously and presents it in a new light, or else goes off in a different direction entirely. It is a bit like the kind of dream in which one finds a door that leads into a new part of a familiar building, or else discovers that one has known how to fly all along, but had never thought to try.

The contributions in this section evoke that quality of creativity and openness. Not necessarily bound to the TSK vision, they express its spirit.

The first article, a "vignette" on knowledge, matches some thoughts on space to the literary form of the playlet. Playfully, it gives character(s) to what can sometimes seem abstract in the TSK vision. The poem that follows, by Ken McKeon, is even more concrete. An evocation of a January day, it invokes knowledge (and thus the TSK vision) only in its last line.

The next selection departs even further from the specifics of the vision, since it was written by someone not familiar with it. The transcript of a talk given by a Zen Roshi who lives and teaches in Canada, it introduces "Being Time," a section of the famed Shobogenzo, by the thirteenth-century Zen master Dogen. TSK students who find themselves looking at time in new ways are naturally drawn to this remarkable text, and we are grateful to the author of this commentary, Anzan Hoshin roshi, for permission to offer this brief introduction, which explores themes that will be inviting to students of the TSK vision. The talk (*teisho*) presented here is the first in a series of eleven devoted to *Being Time.* All eleven are available on the website of the White Wind Zen Community at www.wwzc.org/onlinepub1.htm. The calligraphy that accompanies the talk is also by Anzan Hoshin roshi.

The article that follows, written some time ago by long-time TSK teacher Hal Gurish, draws out the implications of the relation between light and knowledge. The article was written before the publication of *Dynamics of Time and Space,* in which this theme receives extensive treatment. (For a discussion of light in terms of DTS, see the article by Alan Malachowski in *Light of Knowledge.*) Readers may find it interesting to compare the evocative suggestions here with the treatment in these later publications.

The final article in this section treats a recent and popular children's move, *Shrek,* as a parable for the awakening of knowledge as presented in *Love of Knowledge.* At first reading, this may seem more a clever conceit than a contribution to the exploration of the TSK vision, but a more careful reading will show that the comparison is illuminating in both directions.

Where is Knowledge?
A Vignette

Jack Petranker &
Leslie Bradburn

An empty courtyard in the interior of a well-built castle, solid and gray. The ground is covered with cobblestones laid out in an unusual checkered pattern, alternating dark and light. At the exact center of the courtyard is a fountain, but no water is flowing, and it appears that it has been some time since the fountain was in operation.

Enter the King, the Queen, and the Court Jester, deep in conversation.

King: This riddle you have set me is not worthy of you, Jester! The answer is as plain as that ugly nose on your face. Since I am the one who knows, knowing takes place wherever I am. A movable feast of knowing!

He is pleased with himself. Taking hold of the queen, he performs a few dance steps. The queen humors him, but her mind is clearly elsewhere: perhaps still engaged with the question the Jester has posed.

Court Jester: My tongue is down below me but my hair is up above me. My ears are to the right and left. Where am I?

Queen: He's right you know, my Lord. Can you really say the self is located just behind the eyes? How very odd! I doubt you'd find it there if we called the surgeon and had him poke about a bit.

King: Still, that's just how it is. See here! The self is located 'here'. I'm here and you're there. It's all so obvious. The moment I touch my left ear, or even think about it, it's not 'here', it's 'there'. So I have to look for the one 'here' that can't be anywhere else—that can never be 'there'.

Well, that's clear enough, then: Where can 'here' be but right inside my head, just in the middle? And if your friend the surgeon can't find it, it's because his scalpel lacks the right edge. The mind is the only tool sharp enough to find the self!

Court Jester: When I've gone, I've never left. Though you and I will never meet, we've always been at home together. Where am I?

Queen: I do believe he has you, my Lord. Jester! You may be here, but so am I. Isn't my 'here' just as good as yours? And anyway, aren't we both 'here' together?

A table has appeared at the side of the stage. It is made of white metal and is covered with a striped parasol, an anachronism for this time and place. The king and queen seat themselves; the jester sprawls carelessly on the ground next to them and begins massaging his feet luxuriously and with gusto.

Court Jester: Oyez, oyez, oyez! The court is now in session. All who have business before this honorable court, come forward.

While the jester has been speaking, the king and queen have sat with frozen expressions and bodies completely still. Now a bell rings off stage, of the kind used to signal the beginning or end of a meditation session. The conversation resumes.

Queen: I don't see any need for your 'self', anyway. After all, the heart knows how to beat and the eye knows how to see. Perhaps the mind knows how to . . . well, mind!

King: Now that's muddy thinking. The heart beats and the eyes see, but what does it mean to add that they "know how" to do it? No, *I'm* the one who knows how, thank you.

The king and queen freeze once more. The jester jumps to his feet. He reaches into his trousers and pulls out some leather balls, mugging to the audience. He begins to juggle them, perhaps well or perhaps badly. At the same time, he speaks in verse. When the jester speaks, he uses his words to punctuate his juggling, and vice-versa:

A fool can throw his ball into the air.
It's got a part to play: that's all.
And though the king may find it foul or fair
His power shall not make it rise or fall.

The king and queen swing back into action.

Queen: But it does fall! Some knowledge guides it; some pattern determines how things work.

King: Yes of course. A totally predictable event. All objects are loyal subjects, governed by the law of nature. All my life I have seen things fall when dropped. But where's the knowledge in that? The knowledge is in me!

The king reaches into his pocket and pulls out a ball. He aims carefully, then throws it at the jester. He scores a direct hit, and the jester is thrown off balance. The balls he has been juggling fall, and the jester, in trying to catch them, stumbles and falls, elaborately. When he has picked himself up, he walks downstage left and delivers an aside to the audience.

Court Jester: I threw a king off a cliff once. Big he was, and ripe like a peach! And dressed in the finest silks and satins, believe you me! O, He knew how to fall!

King: I heard that! You're just confusing things again. If a king falls, he might as well be a ball!

The jester half attempts to suppress a boisterous guffaw. The king looks uncomfortable. Well, you know what I mean. I can know that I'm falling, but my falling isn't knowing.

Queen: If it's only knowledge when you know it, what is it before it's known?

Court Jester: Yes, yes! Just watch him fall, and before you know it, gravity's at work. We don't need him to teach us that, do we, dearie?

King: I should have known you'd both join forces against me.

Queen: Well, you did know it, didn't you? After all, it's in the script.

TO BE CONTINUED

JANUARY DAY

Ken McKeon

Hey, sweet summer, so far from me,
    Speed up a little, beam more fully, make
    The rich leaves chime in the breeze
    To signify the fog's hold
        Has been broken.
            Let those ample times,
            so negligent till now,
            Assume the crest of the New Year's arc,
            And fall shining on this January day.

Oh muddy world, chilled earth where faces drilled
    In the poverty of need load my eyes up
With the normal dooms:
The gray pains
    Of broken cars, missed buses,
        Ruined shoes, feverish mutterings,

Faltering hearts, gasping lungs,
Or the child's frantic gripping to
The vanished hands of the dead.
Ease up, why don't you?

Enough already! Provide more clearly
In this winter the ways of a home
Where each gesture, able and sure,
Meets the fathering
Anguishings and stumblings that we share,
As true gifts, as sudden guiding stars
That point apart within to bloom
As the inward dawn
Of another's welcoming smile.

Why not that as our natural rambling,
Our meadow's play, our flowering stem,
Both shoot and bloom, even in winter:
The full summer of our ease?

Such questioning speaks to self
And the shudderings we are: I am
A shade among shades,
Knowing only fitfully
The blessings we might be.

May in the course of this year's turn
Knowing come to be more steadily
The source and presence of my life.

Now and Then

Commentary on Dogen zenji's "Being Time"

Ven. Anzan Hoshin roshi

Each and every morning,
first this spring
and now this summer,
a house wren calls out.

He calls out, "I am *this* house wren and this is my house. Stay away."

His house is a hole in the wall of the house neighboring this building. It is a brick wall that faces onto and adjoins the monastery garden.

The house wren sits in this hole sometimes,
and there he calls out his message.

Sometimes, however, he sits amongst the branches
and leaves of the sumac tree.
Sometimes he sits on the wires.

Now he is here
and now he's there.

When here, it is here, now.
When there, it is then.
But when "then" was "now,"
"there" was "here."

Whenever anything happens
it happens now
and "now" *is* this "happening."

Each moment, when it is *this* moment, is right now.

But "then" was once "now" and "right now"
tick tick tick
is now "then."

This moment is not the past moment,
not the future moment

but in this moment
what happens as this moment
both shows and hides the past.

The bird was "there" and now is "here."
The future, however, seems to be utterly hidden.

We can have some very small sense of the future,
in that, for example, if we are walking up the steps
towards the Dokusan room, we can know, somewhat,
the future completion of the motion of lifting,
placing, and setting the foot.

But we might die before the foot
makes contact with the next stair
and go tumbling down.
We don't know. We don't know.

The future is that hidden.

Because it is so hidden from us, we can
engage in all manner of speculation.
We can lose ourselves in
all manner of hopes and fears.
We can imagine future glories,
or complete and abject failure,
or any combination thereof . . . .

And so it can seem to us
that time moves as past, present and future.

Each moment has a past and a future.

Each moment is this moment
and there is only this moment.

But each moment, when it is *this* moment,
is only this moment.
And this moment contains within itself
past and future.

But "past" and "future" only have meaning
when measured from this moment;
and this moment has no width.
It has no depth.

There is nothing that *is* this moment.

There is only this *presencing*,
this activity of Experiencing
that is happening everywhere
right now.

Eihei Dogen zenji says that this is
something very interesting.
Something that we should look into.
Something that we should understand
through our practice directly.

Our practice must take place in this moment.
When else can it take place?

Practicing the realization
which has been transmitted from Buddha to Buddha,
from Awakened Ancestor to Awakened Ancestor,
must take place in this moment.
It must take place *as* this moment.

Calligraphy of "Uji (Being-time)" by Ven. Anzan Hoshin roshi

And when we understand that our practice
realizes itself when we practice not only
in this moment, not only as this moment,
but as the time which unfolds all moments, all times,
then we realize who we are.

And that moment of realization can only happen now,
in *this* moment.

Dogen's essay, his text on this matter,
is called **Uji**: "Being-time."
Ji means "time" or "moment."
U means "Being," or even "has;"
that something has something.
It is the opposite of *Mu*: nothing, no,
don't or doesn't have.

In this text Dogen is not that concerned with
addressing the issue of emptiness, *shunyata*.

Instead, he is speaking about something
further than that, something beyond Mu,
but something which is not a "something."

He is attempting to express something
to each of you
about how Aware Space embodies itself
as each of you
and how each of you unfold yourselves as each other
and as all things, as all beings, all times, all worlds.

And so "Being," in this case, is meant to express
something about the radiance of Aware Space,
the knowingness of Knowing
which unfolds itself as knowns.

Being is always presenting itself as everything that is
and yet it is always absent
because Being can never be "a" being.

Each being
is the presencing of Being itself.

And yet each being is always
becoming absent.
Always vanishing.

In the very moment of its arising,
of its presenting itself,
each being is vanishing.

At the edge of each moment,
on this side the past is gone,
on this side the future has not happened,
and in the center
is nothing at all
because this moment is utterly ungraspable,
is already gone.

And so being-time is Being unfolding itself as beings.
Time is the unfolding of Being as beings.

Being and time are the Activity of Aware Space.
Time is Activity: the radiance of Knowing
expressing itself as spaces which are active
as forms, as beings, as knowns.

Dogen begins his text by raising an old story
from an "Old Buddha," an Awakened Ancestor,
named Yaoshan.

Now, this Zen master Yaoshan
was not always a Zen master.
"Once" he was a Zen student,
"then" he was a Zen master.

Here is a story about Yaoshan when
he didn't know what was going on.
Zen master Shitou said to Yaoshan, "If you say
this is it you've missed it. 'This' and 'not this'
are both way off. What about you?"
Yaoshan said, "Ah. . . ah . . . wha' . . . ah . . . um . . . ."
Yaoshan didn't know what to say.

But now he knows what to say because the Old
Buddha Yaoshan speaks *now* as this *new* Buddha.
But once he has spoken,
he becomes an Old Buddha.
So let's hear what he has to say.

"Being-time stands on the highest mountain peak.
Being-time moves on the floor of the deepest ocean."

Wait a minute!

This isn't what Yaoshan said. Yaoshan said,
"*Sometimes it* stands on the highest
mountain peak. *Sometimes it* stands on the
floor of the deepest ocean."

Dogen has taken the Japanese characters that
can be read as "sometimes it" and reads them
instead as "Being-time."
The "Old" Buddha Yaoshan had something to say,
but Dogen makes it new and has him saying,
"Being-time."

Now what does this mean?

"Being-time stands on the highest mountain peak."
It means that Being-time
looms forth.

"Being-time moves on the floor of the deepest ocean."
What does this mean?
It means beneath everything, here it is.

It also means
up is Being-time,
down is Being-time.
Every place is Being-time.

Up looks down and there is "down."
Down looks up and there is "up."

And when "down" look down,
"down" becomes "up."

Dogen also has Yaoshan say, "Being-time is
three heads, eight arms."
Perhaps this speaks of Avalokiteshvara, the
Bodhisattva of Compassion. The activity of
Awakening beings presents itself everywhere
as everything, in all directions,
embracing everything that is and yet
holding to nothing.

Sometimes Avalokiteshvara has eleven heads,
a thousand arms.
And Aizen Myo-o, one of the five Vidyarajas,
who represents the transformation of lustful patterns
of attention into the expansiveness of
compassionate activity, has three heads
and eight arms.

Yaoshan or Dogen (or both) say, "Being-time is the sixteen foot body of the Awakened One."

(Perhaps I'll read to you from the footnote that I have appended to the text here: "*The text actually says* jorakku-hasshaku *which can mean 'sixteen feet' or 'eight feet'. 'Body of the Awakened One' is a gloss to explicate the passage. The Buddha was ideally represented as towering above the pettiness of common views, understandings, and concerns. His body was radiant and golden. Standing, his form was sixteen feet high, seated, eight feet high.*")

Yaoshan and Dogen go on to say, "Being-time is a staff, a whisk."

B A N G ! *(Roshi strikes his nyoi staff on the hardwood floor.)*

A staff is a symbol of transmission.

A whisk, a hossu, which was originally used to chase away flies, somehow became another symbol of transmission.

Basically a Zen master would hold up his fly swatter and say, "What is this? What is this?"

I guess the fly knew.

"Being-time is a pillar, a lantern."
It is the floor. It is the wall.
Everything that presents itself
within your experience right now
is a moment of being, is Being-time.
And so *you* are Being-time.
"Being-time is Smith, is Jones."
or Zhang and Li, Taro and Jiro.
"Being-time is earth and sky."

Dogen says, "Being-time here means that time,
as it is, is being and all being is all time.
The Awakened One's sixteen-foot high body is time.
Being time has the golden radiance which is time.

Learn that the twenty-four hours of the day are all
the radiance of now. The three heads and eight arms
are time, and because they are time they are
the hours of the day right now."

How is it that 8:15 in the evening
is three heads and eight arms?

What presents itself in one moment
presents itself in all moments.
Each moment, each form,
each activity interpenetrates every other moment,
every other thing, every other activity.

Everything is contained within everything else
but nothing is held on to.
Nothing is jumbled up.
Nothing is obstructed.

Enter completely into this moment
and you will enter not only into
That which contains all time
and expresses itself as all times, all moments,
but you will fall into That
which has no time, no space, no dimension
because it is That in which all dimensions,
all spaces, all times arise.

Within the Total Field of All Possibilities,
the Dharmadhatu, all times have already happened,

are happening, are about to happen.
But the Total Field is simply the totality of
modes of knowing and of knowns.
For Knowing in itself, for who
you each and all truly are,
not only is there no other time than this moment,
there is not even this moment.
There is no time at all.
There is nothing other than Knowing.

And it is now 8:18.

Dogen says, "Without ever being able to really
measure twenty-four hours as long or short,
swift or slow, you still call them twenty-four hours.
The traces of their coming and going are so obvious
that you don't need to doubt them."

Where are the traces of the twenty-four hours?

Our whole lives seem to be littered by
the traces of the twenty-four hours.
You had breakfast. You had lunch and
then you had supper. Not long after lunch
you were looking "forward" to supper.

In each moment,
although the "radiance of now"
shines and blazes
as the vividness of the ungraspable presencing
of That in which everything is arising,

although this moment is the radiance
of your own natures,

and although each moment arises as it is,
as this moment, and does not obstruct
the previous moment or the following moment,

you lose yourself in speculations about the future,
stories about the past.

Measuring yourself by the clock,
measuring time as a kind of passage
in which things come and go
but also extend and endure,
"you" have a sense of time "passing by,"
and you are merely a bystander.

But you *are* time.

You are the unfolding of all moments.

First the bird is here
and then it is there.

And then is now
and there is here.

The continual coming and going,
the rising and falling,
the presencing, the unfolding,
that is your life
is *lived* by you.

Time is not merely something that "passes."

It is the utter and radical impermanence
that makes life possible
so that when you breathe in
you don't have to hold it forever and ever.
You can breathe out.
And when you breathe out
there is room to breathe in.

And this impermanence is so radical
that it must be understood
not as some thing which happens "to" things
but as the activity, the presencing, of Reality.

Joshu Dainen roshi once said something like, "Each moment of our lives is like the hole in a wheel that the axle passes through. The wheel of our whole lives turns around."

Well, actually . . . it would have sounded more like:

"Each momentu . . . you, me, us . . . living is like . . . You know . . . wheel has hole . . . empty hole . . . And something goes through, neh?"

(Someone called out, "Axle!")

"Axle . . . Axle go through hole. Each momentu of our lives like hole. Wheel of our whole life spinning . . . turning around it. Momentu is nothing . . . nothing . . .

gone . . . neh? But too, everything is there . . . only there . . . nothing . . . nothing . . . each momentu say nothing. All nothing of all momentu this momentu . . . neh?"

Each moment of our life
is the total expression
of the Life that lives as all lives, as all beings.

Not only can you not grasp on to it as
"your" life, as "you,"
there is nothing in your life that can be held on to.
Not a single moment.
Not a single breath.

But it is this ungraspability
that presents itself to us as immeasurable richness,
as the gift of our lives,
the gift of this breath,
the gift of bird song.

No matter what the bird might be saying
about territory, about ownership, about breeding,
about hunger, about fear, the bird also *sings*.

The bird flies,
dips and turns and rolls in the air
and plays.

We understand nothing about time
because we understand nothing about this moment.

Dogen says, "The traces of their coming and going are
so obvious that you don't need to doubt them. Yet just
not wondering about them is not the same as truly
understanding them. When sentient beings have
doubts about what they don't understand,
this doubt is usually vague and undefined and
so cannot be the doubt that penetrates
through questioning right now."

Our usual kinds of doubts are all entangled with
struggles against and are dependent upon
our beliefs, our faith in this or that,
our assumptions, our presumptions.
But when doubt becomes questioning
then questioning opens as wonder.

When you realize that you don't know
what anything is
and yet it is still presenting itself,
still shining there—

a drop of sweat running down the side of your face,
the flick of a bird's wing,
the unfolding of a bud into a blossom,
the rising and falling of this moment, this moment,
this moment—

then, knowing that you don't know what anything is
becomes such an open-ended questioning
that it opens into wonder.

And when this wonder is unfolded in
each moment of your life,
not just your "practice," but your life,

when you understand that "practice" is your life,
when you are practising your life,

then the doubt which became questioning,
which became wonder,
becomes understanding.

This is not merely understanding some thing.

You *become* understanding.

Dogen says, "In any case, these doubts were always nothing other than time."

Throughout the Total field
all times are here and now.

In each region and range of the Field,
each presencing arises where it does
and when it does
as a here and now.

But we don't understand.

We think that time moves like an arrow
from the past, through the present, to the future:
You are born. You live. You die.

But we also live time backwards
as if time moves from the future through the present
sweeping everything around us into the past.

Which is it?

In the morning, in the evening,
in the spring, the summer, autumn and winter,
in each moment of your lives
there is only the radiance of now,
the radiance of who you are.

Not at some future point,
not at some point in which you finally
get the point
and understand and realize it.

But even now,
even now,
right now,
with all of your
moments of confusion and
your moments of clarity
you are this moment
of Being-time.

Thoughts Inspired by Light

Hal Gurish

Light reveals knowledge. The spectrum of visible light reveals and validates much of our ordinary knowing. The knowing of insight is also revealed by light, though not light that is ordinarily visible. Given these two kinds of light, we might consider a spectrum of light that is much larger than the spectrum of physical light, an emanation of Great Knowledge present everywhere.

To explore this possibility, consider the widely shared insight that our mental life is closely linked to light. For example, we say that we have a "bright" idea. Then there is the image of a lightbulb going on when we have a 'flash' of understanding. Light also appears in dreams, visualizations, and religious and mystical experiences. Are these examples linked in some way to physical light?

Light visible to the unaided eye is defined by physical properties such as wave frequency. These same physical properties also express its limitations. For instance, light is limited in its ability to communicate by its rate of travel (186,000 miles/second), and it is blocked by opaque surfaces. But the other kind of light is not subject to these limitations. Perhaps light in its larger spectrum does not travel at all, but is present everywhere simultaneously. This more cosmic perspective challenges ordinary knowledge, yet it can also communicate in ordinary terms and language, since it operates within the ordinary as well as beyond it.

Physical light, which reveals knowledge about physical events in our universe, is studied in the physical sciences. Yet it would be wrong to think that 'broad spectrum light' applies only in the humanities and human sciences, since we apply it in great measure to our physical existence. The play of light by various interferences creates surfaces and dimensionality. Ingenious uses of physical light allow us to see hidden worlds. For instance, staining specimens in microscopy produces light interference, which will make visible what could not otherwise be seen. Similarly, different portions of the light spectrum can be filtered to highlight certain features, creating a whole new microscopic landscape that can bring clarity to details of structure and function.

The micro-reality that physical light reveals in this way seems fantastic. Yet the ordinary reality disclosed by the naked eye is no less fantastic. It too has been created by light. As holographic images remind us, the play of light creates appearance, demonstrating that light is inseparable from knowledge.

Viewed from a wider perspective, broad spectrum light and the knowledge it offers manifest as our embodiment in a world that we have created through knowledge. Embodiment seems fixed, but this is an illusion, brought on by making the knowledge revealed in the light of insight into something concrete. The result is to narrow and constrict the insight that would otherwise be available.

We have more latitude in the realm of dreams. In a dream we may seem to return to a world much like the waking world, but this is not so. What appear to be ordinary visual images are generated in a completely different way, not from the senses at all. Nor are they simply reproductions of sensory experience, though that may be one source. The image may be newly created, and it may provide the inspiration to create new forms and meanings that bear the conviction of real knowing. It is not uncommon for us to have a sense of certainty about the significance of our dreams, a sense of knowing that itself bears a light quality, as if something indeterminate has been illuminated.

Beyond dreams is another realm, a no-man's land in which knowing seems to arise without depending on a knower or observer. Looking back after an experience of entering such a realm, we may transpose it into more familiar experience in order to communicate it. A typical image might be to say that it was as if we were standing at the edge of a vastness, looking deeply, and seeing no thing or form.

If we consider that image, the emptiness communicates power and bears knowing. We may not see light, but we cannot quite say that the experience is dark. The knowledge it carries is itself a kind of lightness. Later it may manifest in insight or appropriate action. Here we may be touching the broader spectrum of light. And while this may not happen often, light in its mercurial aspect, light as the messenger of knowledge, could be considered Time—the creator of appearance.

When light creates appearance, it seems too limiting to say that this is a purely physical phenomenon. For instance, an electron microscope relies on a source of illumination smaller than the wave lengths of visible light, yet it give access to images that the naked eye can see. In the broad-spectrum sense, it is light that makes this possible. Can we take this consideration a step further? Could light become so distant from ordinary light, so invisible, as to allow no revelation at all? Could it give rise to a "black hole" of knowledge? We could imagine that at this level, the "absence" of ordinary knowledge might be Great Knowledge—that level of 'in-forming' that is not restricted to existence or non-existence, but includes both.

Great Knowledge includes all possibilities, and in this sense it offers both light and dark. It offers form at any level and all levels simultaneously, awaiting only the dynamic of Great Time to unfold them. In this bounty, generous beyond limits and the restrictions of form, light includes dark, as knowing includes not-knowing. Can we discover this cosmic availability of light within the simple sunlight we all rely on?

Opening Experience to Contemporary Parables

Charles T. Davis, III

In *Love of Knowledge*, Tarthang Tulku (1987) challenges us to "question all beliefs and doctrines" (p. xxvi) and thereby open our inquiry to new discoveries. As a biblical scholar and a teacher, I embrace this process of dynamic questioning, which is basic to the TSK vision and to effective scholarship, and I want to describe an example of its application.

In the first part of this paper, I question the common conception of parables as stories from the past told by revered teachers. I explore a question about parables in my field that is seldom asked, "How did Jesus discover his stories?" I suggest that similar parables surround us. Why can't we see them? Why do we tend to look for our parables in the past? The TSK analysis suggests that the problem may be that we are too locked into our categories to recognize them in the present. In the second part of the paper, I go on to suggest that the movie *Shrek* can be read as a powerful contemporary TSK parable on the need to integrate and transcend the various kinds of knowledge identified in Tarthang Tulku's *Love of Knowledge.*

Parables as a Process of Opening Vision

What is a parable? We tend to think of it as a literary object, but a parable is also a dynamic process using language as a lens for focusing inner experience. This is evident in Huston Smith's (1991, p. 325) comment on Jesus' language in *The World's Religions*:

> A second arresting feature of Jesus' language was its invitational style. . . . Instead of telling people what to do or what to believe, he invited them to see things differently, confident that if they did so their behavior would change accordingly. This called for working with people's imagination more than with their reason or their will. If listeners were to accept his invitation, the place to which they were being invited would have to seem real to them. So, because the reality his hearers were most familiar with consisted of concrete particulars, Jesus began with those particulars.
>
> He spoke of mustard seeds and rocky soil, of servants and masters, of weddings and wine. These specifics gave his teachings an opening ring of

> reality; he was speaking of things that were very much a part of his hearers' worlds . . . . My teachings are true, he said in effect, not because they come from me, or even from God through me, but because (against all conventionality) your own hearts attest to the truth.

Like Tarthang Tulku (LOK, 1987, p. 163), Jesus challenges his hearers to rise above the Tradition of the Fathers, the 'collective knowledge' embodied in convention, dogma, and tradition. Christians very quickly lost this sense of openness and applied allegorical interpretation to the parables (Mark 4) so that an image like "seed" is arbitrarily coded as "the word of God" while the image of the ground is equated with "the heart." Through such abstract interpretation, the images of the parables ceased to be windows opening in the heart and became examples to illustrate Christian teaching.

Typically, readers will ask the well-worn question, "How do we recover the ability to hear Jesus' parables?" The question tends to limit our openness to parables by confining them to the distant past and spiritual teachers. It may not occur to us that we have parables available in our own secular culture. If Smith is correct in his contention that Jesus found his parables embedded in "the reality his hearers were most familiar with," we should ask "Are there similar parables in our popular culture waiting for us to see them and take them to heart?"

The movie producers of our popular culture are our story-tellers. Could some movies be taken to heart just as effectively as the agrarian parables of Jesus? I answer in the affirmative. We only need to challenge our perception that movies are 'just entertainment' to discover them as lenses opening up inner vision to new vistas.

The Movie *Shrek* as a Contemporary TSK Parable

In *Love of Knowledge,* Tarthang Tulku investigates in some detail various kinds of knowledge, each of which in its own way might be considered inauthentic or incomplete. Among these are the modes of collective, descriptive, and intentional knowledge, all of which come into play in the plot of *Shrek*.

Collective knowledge is everything that 'goes without saying' in our culture. It is replete with "routine formulas" (p. 4), cliché expressions, and labels that find a home in "doctrines, traditions, customs, styles, and ideologies" (pp. 44-45).

Descriptive knowledge presupposes a knowing subject who observes and describes the objects of experience. One looks outside oneself and to the past in an attempt to find patterns of coherence so that the world may be ordered through classification and labeling (p. 104).

Intentional knowledge arises as we turn inward to the "subjective realm of feelings and emotions" (p. 156) and look towards the future in an attempt to find personally authentic goals (p. 150).

Technology is a strategy to cope with the disappointments of intentional knowledge. The intentional move inward leads to "dreams and fantasies" that are "tightly bound to disappointment and frustration, or else to dullness, boredom, and anxiety" (p. 158), and we tend to re-value descriptive knowledge as an instrument of power and control. Logic, measurement, rhetoric, and machinery now dominate the search for knowledge. The intentional mode of knowing, however, continues to function unobtrusively in the background, directing our hopes for the fulfillment of our desires into the future (pp. 158–59).

The animated feature movie *Shrek* was released in 2001 with children as its target audience. Our first challenge is to rise above our classification of the movie as 'only for children'. If we can embrace the spirit of play in our approach to the movie, we transcend the collective labeling. We are free to see that the four central characters can serve as pointers to a more inclusive form of knowing. As we allow "freedom and play [to] replace the structures of care and concern" (LOK, , p. 409), Shrek becomes a parable.

The Plot of *Shrek*

The plot challenges our conventional understanding by giving surprising twists to the traditional hero story, with its stages of 'the call to adventure in a perilous realm,' 'the dragon fight that wins the treasure', and 'the return to society with a gift of new spiritual energy' (Campbell, 1973, pp. 36–37). The movie opens conventionally with the potential hero living in a low social estate. Shrek, the ogre, is quite content to live alone in his paradisiacal swamp beyond the borders of civilization. His call to adventure comes when Donkey is driven from society into the swamp, along with other fairytale characters fleeing a pogrom in the City of Duloc. Like the conventional herald announcing the call to adventure, Donkey is a voice of instinctual or transpersonal wisdom. He acts as a wise but irritating adviser to Shrek, always contriving to shake Shrek out of his conventional attitudes and outlooks, especially those regarding personal relationships. Provoked by Donkey and the invading fairytale characters, Shrek grudgingly becomes their champion and goes to the City to seek justice for their cause.

Farquaad is the power-hungry king of the City of Duloc, conducting the pogrom against the fairytale characters. Driven by hubris and a sense of personal inadequacy, he unwittingly falls victim to the Magic Mirror when it advises him that he cannot be a perfect king until he has a queen. Farquaad rejects the 'call to adventure' for himself, and remains safely in the City. He sends Shrek to win his queen from the Dragon—the embodiment of raw, instinctual power—thus unwittingly transferring to Shrek the hero's mantle.

The future queen Fiona, the treasure hard to come by, is locked in a castle ruin guarded by Dragon. She passes her time by reading the fairytale of Sleeping Beauty. Fiona is portrayed as "smart, tough, spirited" and as one harboring a dark secret that shapes her destiny (www.shrek.com).

Romance develops. Dragon falls in love with Donkey, a symbolic union of instinctual wisdom with instinctual power. The conventional 'dragon battle' theme of the hero story thus receives a novel twist. The unanticipated victory through romance replaces the slaying of the dragon theme so that Dragon becomes a magical power assisting the hero.

Shrek still doesn't perceive himself as the hero as he enters the 'hero's journey back into society with the treasure' stage of his journey. He and Fiona fall in love, but the budding romance is threatened by Shrek and Fiona's inability to transcend their personal feelings of inadequacy. It looks as if Fiona will marry the King before he can discover the curse, while Shrek will return to the Swamp deeded to him by the king for his service. But this changes. A reluctant Shrek becomes the true hero who marries the rescued maiden after Donkey persuades him to acknowledge his feelings of love, to disrupt the royal wedding, and to confess his love for Fiona. The King orders Shrek seized and proposes to lock his bride Fiona in a tower forever. At the call of Shrek, the Dragon bursts through the rose window of the cathedral, swallows the king, and intimidates the City.

In the traditional hero story, a socially marginal figure turns out to be the hero who marries the princess and rules her father's kingdom. Shrek gives this a twist. The Princess is transformed. With the kiss of her true love, the curse is lifted from Fiona. She is revealed to be an ogress cursed to live in the social persona of Sleeping Beauty. The movie ends in the Swamp with a raucous celebration of the wedding of Fiona and Shrek. The values of the City are defeated by the Swamp.

Shrek as a TSK Parable

The character Shrek is dominated by the descriptive mode of knowing. As the movie opens, he is reading a fairytale book about a princess held in the Dragon's Keep, waiting for her true love to set her free. Tearing out the page that promises the fair maiden's rescue by a gallant knight, Shrek uses it as toilet paper and gives a shrug that conveys the general sentiment of "fat chance." Nothing in his past experience of the world convinces him that the fairytale could ever come true.

Princess Fiona is dominated by the intentional mode of knowing. Guided by the inner perspective of emotion and feeling, she anticipates a future in which Prince Charming will come as her own true love in exactly the manner promised in the book. The descriptive view would give her little hope since both in the book and in fact many princes have tried to liberate Fiona only to be killed by Dragon. The Princess keeps hope alive by clinging to the vision of a future created by her deepest longing to fulfill the Sleeping Beauty story. Shrek's coming is a psychological doubling of the figure of the unconscious, already represented by Dragon. His arrival does not appear initially to promise Fiona any hope; although Shrek is a much less threatening figure of the unconscious than Dragon. Her future would appear to rest in Shrek's message that she will marry King Farquaad.

Perhaps talk of the unconscious seems out of place in a children's movie, but the movie is overtly psychological. In Shrek's world life is divided between the implicitly circular Swamp, where all is round, soft, and organic, and the City, where all is square, angular and hard—including Lord Farquaad's chin and the square milk bottle. When Shrek and Donkey see the tall square tower rising above the square walls of the City, Shrek wonders aloud in a Freudian vein what a very short King Farquaad is compensating for. There is a delightful exchange between Donkey and Shrek in which Shrek wishes to explore the "layers" of personality only to meet with Donkey's insistence that donkeys don't have layers of depth. Faced with the Dragon's Keep, Donkey confesses that he wears his feelings on his sleeve rather than hiding them below consciousness. As the movie develops Donkey tries to get Shrek to acknowledge consciously his growing feeling of affection for Fiona rather than following a script that says, "Ogres can't marry beautiful princesses."

The names "Shrek" and "Farquaad" are important. Shrek resonates with the German word for "frightening" (schreck), while Farquaad appears to be shorthand for the 'far-off-quadrangle', where all is civilized and artificial, right down to the painted flowers. The visitor entering the City is entertained by a charming animated display that directs attention to the rules. Everything is scripted. There is no spontaneity. Royal officials hold up placards prompting the population to give the proper responses to events. The City is dominated by the psychologically masculine will to power and control. It is a perfectly conscious kingdom in which nothing unanticipated is allowed to happen, since a perfect king exercises absolute control over his subjects. The 'collective view' prevails. The City is driven by a frustrated intentional mode of knowing that has transformed descriptive knowing into a technology that promises to fulfill its frustrated desires.

The City's frustration with the intentional mode is reflected in the pogrom against the fairytale creatures—symbols of our culture deriving from pre-Christian imagination that are considered to be both unchristian and unscientific. Through the subjugation of fairytale, technology prevails over medieval peasant superstition; the rigidity of reason prevails over the fluidity of imagination. The City will no longer fall victim to the hollow promises of the fairytales.

As a frustrated intentional knower, Farquaad cannot completely divorce himself from the magical kingdom. He wishes to possess the technology of the Wicked Queen's Magical Mirror from the Snow White tale in order that it may confirm his judgment that he is the perfect king ruling a perfect kingdom. When the Mirror fails to respond positively, Farquaad threatens to have his executioner smash its glass. The hastily revised interpretation given by the Mirror is that technically Farquaad is not a king since he lacks a queen.

Although Farquaad thinks that he is controlling this captured technology, he is actually being set up. The Mirror shows him three fairytale beauties who are potentially his Queen. Farquaad is seduced by his desires into obeying the Mirror when it asks him to choose one of these beauties from the realm of Fairy to be his queen. The fairytale dimension that Farquaad despised as a flaw in his kingdom has taken charge of his future. In short, the King is possessed by repressed psychological structures. This is an all too common fate for a rigidly rationalistic person or culture. As contemporary history demonstrates, the most elite scientific minds can fall victim to myths without the least sense that they are being controlled by the nonrational forces of the unconscious dimension, the very dimension that they fancy themselves to have banished by their science (Whitmont, 1997, Chapter Eleven).

Farquaad's pogrom of 'fairytale purification' drives the fairytale folk into the Swamp, the unconscious or instinctual world. This arouses Shrek, the Terror that lives benignly below and beyond the City, and he invades. While this horrifies the City's ruler, the population quickly recognizes that Shrek is quite amiable and spontaneously cheer him on in his contest with the knights. Despite conventional fears, the rigid categories of Duloc are 'opened up' as a result of an encounter with the unconscious in the form of Shrek.

The king proclaims that the knight who kills Shrek will become Champion of the Realm. Unfortunately for the king, Shrek wins the tournament in a playful, carnival twist, as suds and mud-wrestling triumph over military arms. For the only time in the movie, the populace of the City claps and cheers without being cued by an official placard. This is a powerful image of how collective knowledge can rob us of our freedom and a sign of how we might be liberated by the very forces that society, through collective knowledge, has negated as the Terror.

Realizing that he cannot kill the victorious Shrek without alienating the population, Farquaad declares him to be the Champion. The conscious, psychologically masculine passion for order and perfection appears to have harnessed the unconscious, psychologically feminine Terror of the Swamp by transforming him into a fighting machine to do the bidding of the King. Shrek must go to awaken Princess Fiona and bring her back to the City to become Farquaad's bride. However, there is a glaring inconsistency in Farquaad's position. If Princess Fiona, a.k.a. Sleeping Beauty, is the soul—the image of the psychologically feminine powers that his kingdom lacks—only Farquaad could hope to rescue her. The point is not wasted on the Princess who threatens to stay in the Dragon's Keep until the King himself comes.

The Dragon's Keep, like the Swamp, is an image of the unconscious, projected as if it were a location in the external world of objects. It is the metaphor for all that society has negated or refused to acknowledge. It is the 'dark side' of reality that society all too quickly tends to judge as exclusively evil. The castle ruins that comprise the Keep suggest the unhappy fate of a culture doomed by that which it denied (the dragon). This is an intimation of Duloc's fate.

Princess Fiona can be taken as an image of the inner feminine potential in every person's unconscious—the lost psyche of the Western world. We first see her in a distance shot that reveals her bed as if it were a stage with the curtains draped for the show in progress. We will learn that she has read the tale of the rescue of Sleeping-Beauty types of princesses and judges her experience by that script. Her submission to intentional knoweldge has trapped her into a very inauthentic life. The Princess is unwittingly imposing upon herself a tightly scripted life. Straightening her dress, feigning sleep, the Princess awaits her knight's awakening kiss.

When Shrek arrives, the Princess is horrified at his insensitivity to the script. He doesn't kiss her; he hasn't killed Dragon prior to waking her; he isn't her true love but only the King's delivery boy. At this stage in the story, Princess Fiona is a persona—the mask with which we meet the world. She isn't yet the authentic image of the psyche/soul-process. She is a fake front for an authentic but hidden identity that only emerges at sundown as an ogress. Contained in the intentional mode of knowing, the Princess has no sense that she is living a lie. We might say that she has an inverted sense of reality, in which the fake appears as real while the authentic appears as the fake.

In our persona-mode, we follow the script that society gives. We project a fake soul-image that pleases the guardians of collective knowledge. Living by the book and the 'read-outs' of collective knowledge rather than from our capacity for self-knowledge, we are prisoners of our failure to develop a love for knowledge. Only as we begin to question the 'unquestionable truth that goes without saying' can we hope to claim our freedom to live authentically—a lesson Fiona will soon learn.

Princess Fiona is shocked to discover she has been rescued by an ogre—an inner man-image, or animus, both ugly and lacking all the civilizing accoutrements listed in the script for the true love. On the journey back to the City, the action follows the plot of Beauty and the Beast. Shrek falls in love with the Princess despite her beauty; she falls for Shrek despite his bestial features. Through the love that develops quite unconsciously as a response to the deeper dimensions of the person, Shrek and the Princess are transformed, momentarily freed of their conscious agendas.

But such momentary awareness is not enough to arouse courage for a radical questioning of the script. Despite Donkey's urgings and their own sense of a developing love, Fiona and Shrek follow through on their previously scripted plans. Their own sense of inadequacy leads them both into a misunderstanding. Fiona is anxious about how Shrek will respond when he discovers that she is ugly by night. As she confesses this to Donkey, Shrek overhears only part of the comment and is deeply wounded by the thought that Fiona regards him as ugly and inadequate. This misunderstood comment confirms a fear that he harbors. Neither Fiona nor Shrek has the courage to confess their love to the other, to Donkey's great consternation and despite his efforts to effect an honest encounter between the two lovers.

Shrek, the ogre, and Fiona, the princess, are complementary. From the perspective of descriptive knowledge, Shrek sees no value in following the fairytale script and is able to liberate the Princess from her imprisonment by the text. By insisting upon the value of the book, the Princess opens Shrek's eyes to the possibility of wonder and happy endings. By the end of the movie, Shrek and the Princess will have come to experience something like "global knowledge" (LOK, p. 408):

> Attuned to conventional knowledge without accepting its assertions, a global knowledge would accept propositions as if they were true—either for the practical purposes of our lives, or for the specific purpose of cultivating greater knowledge.
>
> . . . . Within the luminous vitality of such a knowing, existence would be transformed into 'appearance as'. 'Was' would be 'as' and 'is' would be 'as', essence would be transformed into absence. Negation would be present within all appearance, as the heart of appearance.

Donkey prods Shrek out of his denial of his feelings for Fiona, prompting him to disrupt the loveless wedding by confessing his love for Fiona. With the aid of Dragon, they arrive at the cathedral in time. Through the agency of Donkey and Dragon, care and concern will eventually give way to ecstasy in play and celebration for both Shrek and the Princess.

Fiona is cursed by a witch, as in Sleeping Beauty. The prophecy is that "when you receive your true lover's first kiss, you will take love's true form." Fiona, who is going by the book, interprets the curse conventionally: she will be her true self, Miss Perfect Princess, for her curse is surely to assume the form of the ogress at night. Her situation presents an apt image for how we lay aside our persona in the privacy of our bedroom, where frightening dreams may come to challenge all of our daylight certainty. When Shrek's loving kiss lifts the curse, reality appears right side up, and to her total dismay, she discovers her true nature—an ogress cursed to identify with a Perfect Princess persona. She and Shrek leave the wedding in the Swamp in a Cinderella-type garlic carriage amidst a raucous celebration that stands in vivid contrast to the orchestrated rituals of the City of Duloc.

The movie ends with a clue that it can be perceived as a parable of liberation from limited ways of knowing. When the movie opens, the fairytale book bears no title on its cover. Shrek regards it as garbage. At the end of the movie, the book bears the title "Shrek." The fairytale of the Sleeping Beauty has become Shrek's story, his unintended heroic journey. As a result of his loving relationship with the Princess, Shrek has negated his original negation of the happy ending promised by fairytale, even as the Princess has negated her rejection of life in favor of following a script.

Conclusion

From the perspective of cultural criticism, Shrek is a parable of our need to restore the psychologically feminine to our technologically focused culture. The wonder of life as portrayed in fairytales needs to be recovered. Imagination is needed to supplement intellection. Above all, we need to challenge the scripts that we cherish even as we hold them to be worthy guides, lest we purchase a beautiful mask (persona) at the price of an authentic self.

REFERENCES

Campbell, J., 1973. Hero with a Thousand Faces 3. Princeton, NJ: Princeton University Press.

Smith, H., 1991. *The World's Religions*. New York: HarperCollins.

Whitmont, E., 1997. *Return of the Goddess*. New York: Continuum.

EXHIBITIONS

In this section, we consider various fields in which the TSK vision could offer new insights and approaches: arenas for knowledge to exhibit itself in space and time. Of course, the range of such possible fields is unlimited, but in the years in which TSK has been introduced, it has proved of special interest in psychology, education, and business, each of which is represented here.

The first article, by Victoria Riskin, sets out in a preliminary way the distinction between a standard psychological approach, centered on the self, and a TSK approach, which treats 'mental health' as a concept that does not have to be bound to the idea of a self. Several authors writing in the field of psychotherapy or personal transformation have relied explicitly on TSK, but as yet no practice based directly on TSK has developed. Is this simply because TSK is not well known, or is there a point at which the tension between a self-centered approach and an approach that looks beyond the self leads to a split?

To go further, readers may want compare the distinction made by Meindert Gijzen (a practicing psychologist) between "Inner Technologies of Freedom" and the TSK vision, available in his contribution to *A New Way of Being*, the companion volume to this one.

In his article, Ken McKeon discusses the application of TSK to teaching. A glance at the list of contributors to this volume will show that TSK has a special appeal to individuals with an academic background. Here, however, the author is concerned with the way that TSK can change experience in the classroom. This is an area that seems specially suited to TSK, which points out repeatedly that the patterns that block access to new forms of knowledge are laid down in layers during the course of our education and acculturation. Until now, however, it has received relatively little attention. The present article offers some hints as to how it might be possible to take some initial steps in this direction. It would be deeply interesting to know how teachers who expand on these suggestions fare, and to listen in as they compared notes.

The final three articles in this section all relate to business. Interestingly, it has been in the fields of management, business consulting, and organizational development and transformation that TSK has found the most ready access. Perhaps this is because the world of business is less concerned with the pedigree of an idea than with whether it has a measurable impact. Managers have always been interested in the way that time affects the managerial process and the bottom line, so this is one natural place for a connection to be made. It may also be helpful that Tarthang Tulku himself has addressed issues centered on work and management in at least two books, *Skillful Means* and *Mastering Successful Work*. The latter in particular contains extensive discussions on the nature of time and the opportunities it affords for transformation.

In the first of the articles, Robert Reyerse, a business consultant, moves gently from his own reflections on the possibilities for a TSK kind of knowledge into the ways that this possibility takes on substance in the practical world of business. His special focus is the resistance to change. His analysis of what makes change so difficult sheds light on recurring themes in the TSK books (for instance, how the self structures and relies on confusion). And his attempts to implement TSK in a specific consulting context suggest ways to translate the TSK vision into business realities.

The next article, by Ronald Purser, makes the connection between our views of time and our ways of interacting with the environment, especially as corporate and organizational actors. Here is a field with enormous potential impact. If TSK could activate a different understanding of how we interact with our world, it would perform a service for all humanity.

The final article in this section, by Steve Randall, grows out a program in time management for managers and others; much of it is available in different form on the web at www.manage-time.com. Randall distinguishes between "conventional time management" and "inner time management." The latter, he writes, must proceed on the basis of rethinking the nature of time, a step that ordinary time management approaches would probably consider beside the point. Although this article does not refer specifically to TSK, anyone familiar with the TSK approach will find much that sounds familiar, as well as a whole new set of practical issues and concerns to reflect upon.

Tapestries of Healing

Victoria Riskin

How do we relieve suffering? How can we lift ourselves out of confusion and conflict and live a life of fulfillment? This issue, which has challenged human beings for thousands of years, has preoccupied modern psychology since its inception.

The psychological perspective posits that human suffering stems from the individual's interaction with the environment in early life. Maladaptive patterns established at that time persist into adulthood, causing continuing and recurring difficulties.

In its classic Freudian formulation, this view saw human nature as basically driven by untamed "id impulses" that could only be contained through socialization and injunctions imposed by parent figures. These injunctions were internalized in the "superego." Id and superego engaged in conflict, with the ego the arbiter in their struggle. This view gave psychology a specific task: to help to bring to consciousness forgotten feelings, memories, or impulses and help the patient accept them, understand their neurotic source, and make a healthier adaptation.

Today many psychologists consider this view outdated, yet the fundamental orientation it established still permeates modern psychology and indeed all of Western culture. Almost all therapies, including those practiced by learning theorists and body therapists, agree that the source of present difficulties is conditioning caused in the past, with a special focus on the parent-child relationship as the primary source of this conditioning.

This past-centered view of an "historical self" has become so much a part of contemporary thinking that we hardly notice it in operation. Operating within that view, we cannot see its impact, but it guides our vision. Indeed, alternative views from other cultures (for example, that strange behavior or suffering is caused by demons, witches, or evil spirits) are dismissed as childish and unenlightened.

Psychology's commitment to the historical self may place severe limits on the potential scope of treatment and healing. Bound to the past, the self becomes solid and tenacious at the core, with change possible only at the fringes of its solid structure. How can we even investigate the root causes of suffering if the self constantly asserts that it is as it is; if it constantly defines and labels experience on the basis of the past? Unless we are willing to question the claims and views of the self, and perhaps even its existence, we will be confined to adapting our perspective to entrenched patterns and a dominant world view.

It is as if we were located in the center of an elaborate tapestry in which the self represents only one small knot. The tapestry has pattern, color, and texture, but since we are caught up in our narrow, dense knot, we miss the larger pattern, and even the pattern that we make within that larger pattern.

The structure of past-present-future centered on the historical self may well confine our vision and close off appreciation for the whole. This focus on the historical self is not wrong, but it errs in mistaking a small part for the whole. Conventional therapy works within this structure. It helps the individual blend more effectively into the fabric, thus reducing tension around the edges and making more satisfying connections. But there is no attempt to open to the whole or change the position the self appears to occupy within the fabric.

The TSK vision suggests that it is possible to see the fabric as a whole, while understanding the arising of the self as part of a more all-encompassing patterning (TSK, p. 37):

> When we open all our perspectives and our senses and learn to view life in a holistic way, we can see that the time, space, and knowledge which have established restrictive perimeters have done so only because they have been insufficiently challenged, explored, and appreciated.

If it were possible to shed light on the macroscopic and microscopic patternings of the tapestry—the models and stories told by the self, established by the culture, and presented by linear time itself, what results could be expected? According to TSK (p. xxxix):

> As our perception of time and space opens, we can begin to see and appreciate a . . . knowledge beyond duality and dichotomies, which is all-pervasive. With this knowledge we can learn to see the playfulness of all situations; we can open to the natural tendency of our energies to flow toward the exploration of understanding and beauty—toward the deeper values of humanity . . . . We can then directly experience our Being, which expresses itself as a dynamic and complete freedom. In this way we can discover what it means to be truly human.

Can the therapist make use of this vision to promote a deep healing? How can TSK be used in the domain of therapy, and for whom is it appropriate? Many questions remain to be explored. What is clear is that the vision offers innumerable starting points for the development of a deeper knowledge.

TEACHING AND THE TSK VISION

Ken McKeon

A number of years ago I bumped into an old friend with whom I had gone to graduate school at the University of California. He asked me what I was up to. I answered that I was teaching English in a high school and was also studying and teaching Time, Space, and Knowledge. His response to this last news was an immediate rhetorical question: "What else is there?"

My friend had caught a taste of the sweeping comprehensiveness of Tarthang Tulku's vision and its intuitive appropriateness for our age. This is an age grounded in science and disbelief and suspicious of all hierarchies, with their imposed but no longer imposing dogmas. It is also an age in which, for some, limitations, when noted and engaged, provide the occasion for playful inquiry and a lively and direct activation of the human capacity for appreciative knowing.

My friend, to instance one such an occasion, might have activated his rhetorical response, a kind of response which automatically seeks to impose its own form of limitation, by asking a further question: "So how do Time, Space, and Knowledge appear for you, play out for you, in your work as an educator? Are they simply names for concepts that fuzz out particular situations by making them pretentious, and thus deadening? Or are they of any use? These questions were not asked, but had they been, here is the response I would have given.

To begin with, TSK is conceptually rich and wonderfully adaptable. It claims no truths and establishes no ideological frameworks or systems of belief. Like metaphor, the most playful power of language, TSK is transformative and illuminating. It opens up perspectives that lead one within, to open and lively circumstances that are encouraging, not restraining.

For example, consider a primary TSK concept and practice of noting and opening up the boundaries that define (and so limit or freeze) ourselves and our experience of the world. By opening, penetrating, or thawing those boundaries we and our experience also open, and the whole set of things that presents becomes more interesting and engaging.

Now think of teaching a piece of literature, Thoreau's *Walden* for instance. Students often find Thoreau something of an obnoxious ranter as he rails against the materialism of his times. As in almost any situation, such students are onto something: a particular is defined and limited by its own defining, and so are the particulars pointed to by Thoreau. But as always, there are further steps to take.

One could simply ask these students and oneself to imagine Thoreau as being walled in by his rants, as being defined by them, as are those Thoreau points to, those caught in their particular ways of being, as seen by Thoreau. At this point questions pop up. What activated the rant? Who responded to it? With what motivation? And what of the students themselves (and their teacher)—how do they respond to it? What activated their response? Of each responding element, citizens and author and class of students, one could ask: What sort of space or room do they find themselves inhabiting? What are they being kept from: what qualities of experience, what ways, natural or unnatural? What avenues of knowledge or experience . . . ?

Various students, like their teacher, will probably adopt positions or walls taken up in response to the language of *Walden* and the situation occasioned by that language. Those positions can be described with care, can be inquired into, can be momentarily embodied or discarded, can be seen as pointing to contemporary concerns . . . . The alternatives surge, and so does the classroom. The arising energies can be seen as qualities of time, time that can be ridden like a wave.

We have seen a particular space presented (the referents and meanings established by the text), and an activation of time (the various positions taken by the individuals involved). A sense of appreciation can be brought in here, a sense (in TSK terms) of knowledge. Simply encouraging those involved (again including the teacher) to pay attention to the qualities of awareness as it mirrors their experiences; as, that is, experience in all its aspects is given as knowledge.

As knowledge, or knowing, presents itself more fully, a shift occurs. In the nutshell of this imagined class, the encouraging promise of TSK is being realized. The nut is cracked open, and the meat within the opened shell is found to be nourishing and sweet.

In the above scenario, the process went from the abstract and conceptual to the practicalities of the classroom to some state approaching an active embodiment of the vision. The initial impulse was not to get to some special state, nor was the instruction ideological, moralizing, or spiritual, although those categories could be brought into the mix as well, for nothing is necessarily excluded. The basic impulse was to prompt appreciative inquiry, an inquiry whose presence is signaled by a quality of ongoing meaningfulness. Or to put this in more low-key terms, the inquiry is signaled by a quality of having fun.

So to return to my friend's question about what I was up to, the responsible answer well might be: I seem to be engaged on a regular basis, as a teacher and a student, in having fun. That the fun has been inspired by a vision—that it seems to be professionally, educationally, and personally productive and satisfying while also transcending those terms—might well cause us to burst out in laughter.

Is TSK Practical? An Inquiry in Business Consulting

Robert Reyerse

The investigation of human experience certainly predates recorded history. Still, it is the surviving words of individuals such as Socrates, Plato, Confucius, Buddha, Descartes, among many others, that provide us with a limited appreciation for the depth, breadth and variety of inquiry that has taken place over time.

What marks these thinkers, their questions or their philosophies, as important? What did they contribute to the evolution of inquiry that makes them memorable? Perhaps it is the new paradigms their questions ushered in, or the new conceptual models they developed. It could be the influence their questions have had on human social structure and behavior. It could be their very approach to inquiry or their courage in asking the difficult questions. What seems consistent is that great thinkers are remembered and respected as much for their questions as their answers.

Some may find the great questions awe-inspiring, but others will probably ask, "So what?" The average Joe (in my case, my friend Roger) will wonder what all this means on a practical level. Roger thought it over, and reached his conclusion: Very little.

My answer to the "so what" question is straightforward. The great thinkers have impacted our lives in myriad practical ways. Human values, government institutions, approaches to education, and scientific investigation all bear the imprint of the mental footsteps of some of the world's greatest minds. The investigation of human experience through time has shaped our current world. More important, it has shaped the assumptions and unconscious models that we use to assess and judge the world our senses perceive.

At this point, the conversation turns to the Time, Space, and Knowledge (TSK) vision. Challenging these forgotten and hidden assumptions is where TSK has taken on a particularly important role.

"Okay," says Roger, "Tell me how it works." Now I have a problem. The TSK approach to inquiry resists any simple description. More than just an open mind, it asks us to have an empty mind—at least, a mind empty of assumptions, preconceptions, scripted methodologies, and any other obstruction resulting from what we think we already know. As Tarthang Tulku says (DTS 16):

> I have often been asked how TSK could be classified as a field of study. The question has no easy answer. It is not philosophy or psychology and not religion. Perhaps it could be seen as an adventure of the mind, with the potential to branch out in many directions. For someone knowledgeable in a particular field, TSK may be able to inspire new ideas related to that discipline. For others, TSK may stimulate clarity and a sense of value.
>
> My own view is that TSK should not be considered its own discipline, at least in the sense that it conflicts with other approaches to knowledge. I see no contradiction between TSK and my own Buddhist background, just as I see no conflict between TSK and any other tradition of knowledge.

Put in more positive terms, TSK is about a dynamic form of inquiry that focuses on questions:

> Answers are not the purpose of our questioning. When we learn how to ask fundamental questions in ways that are fresh and alive, we conduct into our lives an intelligence that applies directly to our own immediate circumstances. In activating this kind of inquiry, we can rely on the great masters and thinkers of the past for inspiration and guidance, but their answers cannot be our answers. We must each individually take up the challenge of knowledge for ourselves. (VOK, p. ix)

Still, I am painfully aware that this is all quite abstract. Roger reacts skeptically. He sees the TSK vision as an esoteric approach of limited value, good only for intellectuals with time on their hands. He puts this challenge in the form of a simple observation: "When I walk through the forest I am surrounded by trees. I am consoled and overjoyed by the simple fact that 'a tree is a tree'! And no amount of TSK questioning, mind expanding game playing, or drug-induced altered states of consciousness will change that fact."

This challenge took me aback. Roger's "tree is a tree" haunted me for several days. He seemed to be saying that a tree, or any object, for that matter, is nothing more than what it appears to be. He seemed to suggest that investigating basic understanding or perception is not practical. Since I was planning to write an article about the practical application of TSK in a consulting environment, this challenge made me nervous.

I walked around with an active internal dialogue on all this for a few more days. Then, perhaps as a result of too many TSK exercises (or some minor psychosis), a new voice from somewhere deep inside asked, "Is a tree just a tree? And if it is, what is it?"

"Is a tree just a tree?" I decided to investigate. If a tree was just a tree, not worthy of further analysis, then investigation should have no impact on my understanding or appreciation of the tree. Well, I would see.

I started with the dictionary definition (Webster's On-Line Dictionary): "a perennial woody plant having a main trunk and usually a distinct crown." This seemed less than helpful, so I turned next to my own limited knowledge of trees. Trees are plants that grow from some sort of seed. They come in a variety of sizes. They can be coniferous or deciduous. They tend to be green. Also not too helpful.

It was time to take an TSK approach. I went outside and sat down by the incredibly huge, hundred-year-old cedar in my backyard. In preparation, I did my best to discard whatever preconceptions I had hidden away regarding trees. My plan was to proceed by simply asking (and as far as possible answering) basic questions.

I decided to start with questions relating to time. At this present moment the tree appeared to be pretty much what the dictionary had described. What had it been before it was a tree? At some point in time it had of course been a seed. Somehow combining that seed with water, dirt, nutrients, and sunlight had turned that seed into a mighty cedar. How had this feat been accomplished? The seed contained or had acquired knowledge of what was needed to create a cedar, space allowed the ingredients to be available and gave the tree a place to be, and time allowed the dynamic process of growth to take place. Looking at the tree as a process over time, a tree was the localized knowledge of how to transform one group of materials into another. A tree could be seen to be a living transformation machine.

Focusing on this ongoing process of transformation with an ecological perspective, I saw that this tree in front of me was not only a transformation machine, but also a recycling machine. On an ongoing basis it recycles sunlight, carbon dioxide, water, and various nutrients into wood, oxygen, evaporated water, leaves, seeds. What makes the recycling theme especially compelling is that the nutrients quite likely came from a previous tree.

Sticking with the theme of time I pondered, "What about when the tree is no more? What will this tree be after it is no longer alive?" Taking a relatively short-term view, and assuming no human intervention or natural disaster, I could expect that the tree at the end of its life would fall down, decompose and form the nutrients for bugs and the next generation of living plants—a continuation of the recycling process. If humans did get involved, the tree would most likely be harvested, turned into lumber that could be used for a variety of purposes, from building a house or fence to making furniture. So a tree could also be seen as potentially a house, a fence, or a chair.

But there were also other possibilities. The tree could be struck by lightning and burn up, transformed into heat, smoke and ash. A different form of human intervention could result in the tree being cut down and turned into firewood. This was interesting: a tree might turn out to be the fuel used to warm the house made of another tree.

Taking a pause, I had to wonder, "How did humans make this great discovery that trees, apart from being trees, could also be furniture, fuel or houses?" What person sitting in front of a cedar decided to question the uses of that tree? I could imagine possibilities. Watching a tree burn after being hit by lightning could certainly put the idea into one's head that smaller pieces of the tree could accomplish the same miracle of fire in a much more controlled manner. Seeing a tree float down the river might just convince someone to hollow out an old fallen tree and turn it into a dugout canoe. Still, who first thought of binding a few pieces of wood together to form a chair or a bed? Who came up with the idea of stacking logs to build a house? Seen from a perspective of a complete unknowing of what a tree could be, these were incredible insights: bursts of knowledge at its most practical.

With knowledge in mind, I returned to my backyard cedar. I had already determined that this tree held the blueprint for making a cedar inside it, passed on presumably from the seed from which it sprang. Setting aside the obvious question of how the information came to be in that seed, I wondered, "What other knowledge does this old tree contain?"

I began mentally disaggregating the tree into its major constituent elements. The leaves (more like scales) of the cedar had the knowledge for photosynthesis, the process by which the tree converts carbon dioxide and the energy of the sun into some form of carbohydrate, and (happily for humans) oxygen. A tree was not simply a tree: it was knowledge embodied in the form of a chemical factory.

Looking at the bark and the woody interior of the cedar led to the same conclusion. The tree's wood is heavily soaked in a resin that is a natural bug repellent. Clearly cedars know how to create a chemical that serves to repel their potential bug predators.

I looked again, from a different perspective. Everyone knows that a tree grows in rings that record its seasons of growth. The rings store information on everything that the tree has experienced and endured, from the weather to major ecological disasters. So a tree is a clock and a window into the past.

Finally, I took a space approach to the tree. Taking on the size of a fly I circumnavigated the immense base of the tree and then flew to the top, some 100 feet above. I entered an immense world, sheltered from the sun and wind, populated by a huge variety of insects, a few squirrels, and birds of several different species, and festooned with moss and lichen. This was not just a tree; it was a complete ecosystem!

After an hour I brought my exercise to an end. I had only scratched the surface, but I had managed to draw two important provisional conclusions:

☆ A tree is not just a tree. More precisely, the use of language to create a simple, stable model of reality ("this is a tree") may be useful, but it carries with it the danger of overlooking the rich diversity of what experience presents.

☆ TSK is a powerful approach to analyzing concrete practical questions.

This preliminary exercise renewed my confidence that TSK can be practical. Now was the time to put that feeling to the test. I turned to the original theme of this article: TSK and Consulting.

TSK and Consulting

I have been actively involved in consulting for some twenty years. I have undertaken a wide variety of consulting assignments ranging from strategic planning to designing management information systems to business-process redesign. My first ten years were spent working for a large international consulting firm, where standard methodologies were employed to ensure a standard and consistent result. In the last ten years I have worked as a sole practitioner, which gives me the opportunity to develop my own approach to particular assignments. This has let me experiment with using TSK approaches in a consulting environment.

Throughout my consulting career a number of issues or obstacles have remained fairly constant. In general, people:

- ☆ do not like change;
- ☆ have difficulty letting go of their own ideas or entrenched positions;
- ☆ find it hard to be creative;
- ☆ cannot readily see from more than one perspective;
- ☆ have difficulty seeing the implications of organizational or process changes;
- ☆ cannot readily identify their own impact on the dynamics of a project or process.

This is certainly not an exhaustive list of issues, nor is it carefully formulated on the basis of a rigorous academic investigation. I offer it simply as a personal and subjective list of issues that provide a practical, real-life laboratory for testing out TSK-based techniques. What unites it is that all these issues are related to people's perceptions, attitudes and ideas.

Consulting firms and business-school academics have spent a great amount of time and money developing methodologies to deal with these and similar issues. Approaches such as change management, change readiness assessments, knowledge management, organizational culture assessments, creativity workshops, effective leadership programs, and performance-based personal planning are examples of approaches that have been tried with varying degrees of success. In my experience, the challenges remain formidable. For instance, implementing a knowledge-management system will not guarantee that people will share their knowledge if the culture of the organization has historically discouraged sharing. No one approach can readily deal with these interacting issues.

The idea of using TSK techniques to address these issues came to me when I was reading *Dynamics of Time and Space*. Two quotes captured my attention.

> Our present tendency is to understand the future as an extension of the pre-recorded structures of the past. (DTS, p. 92)

> The possibility offers itself for a knowing that does not follow the old ways, the 'has been taught' way of the predetermined automatic. (DTS, p. xxx)

The first quote rang true to my own experience, which is that many people seem stuck in the familiar rut of "we do it that way because that's the way we have always done it." It was also particularly relevant to many of the planning exercises I had been involved with or witnessed, where analysis of the past was used as the basis for predicting the future. The second quote offered hope for an alternative way of knowing. In 'consultese', TSK seemed to be offering a way of "thinking outside of the box."

At this point I focused on the first two issues on the list above, as they were very much related: avoiding change and holding on to entrenched ideas and positions are really two sides of the same coin. Could TSK provide observations, insights, or techniques to help overcome these obstacles?

I created a worksheet for these two obstacles and assembled a few relevant TSK observations. The table on the next page provides one version of the result.

It seemed clear that TSK understood the issue of change and the problem of letting go of entrenched positions. I decided to investigate further.

A TSK Approach to Change

Attitudes toward change naturally span a spectrum. For instance, younger people seem more open to change than older people. Those with little to lose or much to gain from a change in the existing structures are also more open to change. Still, the desire to avoid

Issues	TSK quotes
People do not like change.	From our conventional perspective, the events and facts that we know—the tremendous weight of our past and of our cultural conditioning—have seemed to establish a vastly complex world within which our present positions gain their meaning. (TSK, p. 212) Other alternatives may present themselves—other ways that things could be, other conducts we could choose—but we know in advance that they are not worthy of serious attention. (DTS, p. 122)
People have difficulty letting go of their own ideas or entrenched positions.	Another danger comes when we let ourselves imagine that our specific values and understandings are right, and that we have nothing to learn from others" (SDTS, p. 142) When we adopt a specific identity, we inevitably adopt its boundaries and limits as well. (DTS, p. 26)

change seems the norm. And this applies even when dealing with business processes that are clearly seen as sub-optimal or unacceptable, and where all involved parties agree change is required.

For instance, I recently had an assignment to implement a new management-reporting system. Initially all stakeholders were enthusiastic at the idea of change, but when the time to start the development process arrived many people had second thoughts, and took positions that undermined the process. They came up with reasons why the *reports* could change, but what *they* did should stay more or less the same. They proposed new activities that were essentially the same as the old, repackaged in some non-essential way. They came up with a multitude of reasons why the new processes wouldn't work. The desire to keep things as they were gained the upper hand.

Love of Knowledge (p. 224) proposes an interesting hypothesis for explaining this phenomenon:

> Another response to the momentary awareness that matters could be different is to interpret such awareness as the onset of confusion. Confusion is the self's response to the not-knowing that calls a story into question—a way of twisting not-knowing to gain control over it. The self expects to understand in terms of the old categories and old stories and interprets the failure of the old as a failure of understanding: a sign of weakness that it quickly turns into a shield against the new. Aware that the old

patterns have been brought under suspicion, it holds onto them as tightly as ever, catching awareness 'in a bind'.

My own related perception is that change moves people from a situation that they know to one that they do not know, or from activities they know they can perform to processes they have not yet performed. The resulting fear of the unknown or fear of failure may be vague, but it seems psychologically important.

Could TSK offer any practical solutions to this problem? To answer this question, I developed a list of possible TSK solutions with the plan of testing them on a upcoming project, one that involved developing a new "finance function" at a Global Asset Management Branch of one of the world's largest banks. My approach was to take an TSK technique or idea and translate it into a business-friendly approach and vocabulary (see facing page).

Results

At the outset it must be stated that this was not a scientific experiment with a test group, a control group, and a structured approach. It was simply my personal application of a number of specific techniques employed on an *ad hoc*, subjective, and often spontaneous basis. Despite these methodological flaws, however, I think the results are interesting enough to be worth sharing.

TSK Approach	Business-friendly Approach
Inquiry dissolves boundaries "Inquiry restores the untrammelled energy within appearance. Challenging us to question the structures of reality, it shows that we cannot rely on 'this' and 'that', on 'yes' and 'no', on 'something' and 'nothing', on 'past' and 'future'. As these distinctions lose their significance, we no longer know our limits or limit our knowledge." (DTS, p. 65)	**Inquiry is the answer** • Have the project team keep asking how things could be better or different. • Inquiry lets the team step away from known patterns and structures. • Keep the inquiry going: no assumption is left unexposed and no solution left unproposed.
Let go of a particular position "We can develop a mode of 'seeing' *which is not limited to a particular position or 'point of view'* at all." (TSK, p. 27)	**Let go of your own ideas and approaches** • embrace ideas of others • be aware of bias against ideas of others • be aware of holding on to your own idea and how this shuts out openness to new ideas
Speculate "One alternative to the standard ways of knowing is speculation." (SDTS, p. 147)	**Generate ideas** • forget about how things are now • let go of projecting past into the future • let go of what we know or think we know • focus on an ideal situation, regardless of its possibility or how it could be reached through time

Inquiry dissolves boundaries

In developing the new finance organization, I was instructed to disregard "what was" and create "what was required." I decided to employ brainstorming sessions, a standard tool in the development of any new process or organization. Incorporating a TSK-like approach in these brainstorming sessions resulted in extremely effective meetings and results.

At the outset I explained to participants that we would take a question-based approach and that no question would be considered off-base or irrelevant. Initially, we questioned what the organization as a whole needed from Finance, to identify what key processes were required. As basic requirements were identified, these too were questioned. Why were they required? How were they required? When were they required? By whom were they required? What quality was required? Was Finance really the right group to meet these requirements? The result was a list of process requirements that were developed very quickly and effectively, and in my subjective opinion had unusual clarity.

What was perhaps most interesting about this process was that the team member who was able to pose the most questions, whose approach to inquiry was the most open and dynamic, was also the person most enthusiastic about the possible changes. This remained the case after implementation. I am not sure of the relation of cause and effect here: Was this a result of the approach, the nature of the person, or a combination of the two? I cannot say.

Let go of a particular position

This particular practice had personal relevance. I have found that once I develop a new process, I am not particularly open to alternatives or even to constructive criticism. Over the years I have made some progress in this area, but I know that team members still sometimes express frustration that I have not given them the opportunity to contribute. So I decided to carry out this experiment for myself. I used two simple, complementary rules:

- ☆ I would embrace ideas presented by others as if they were my own.
- ☆ I would treat my own ideas as if they came from someone else.

This exercise led to challenging insights. I saw immediately that the value I placed on a new idea was very much influenced by the person presenting it. Of course this was not a new insight, but the exercise brought it uncomfortably into focus. My own ideas had a special feel that I could not let go of: not just ownership, but a surrounding feeling of pride. As for ideas presented by others, when I respected or liked the person it was much easier to embrace his or her ideas. I made an effort to compensate for this bias, carefully dissociating the idea from the person as far as I was able. This seemed a reasonably effective approach. Others reported similar experiences, and in the group as a whole, all team members, regardless of personality issues, reported that they had felt able to contribute to the process.

Speculate

The trap of projecting the past into the future, described above, is very common in developing new processes or new organizations. One reason is fear of change, but another may simply be lack of creativity. I felt that TSK speculation, and especially letting go of the constraints of time, could be very helpful here.

I decided to test this approach in the task of developing the vision for the new Finance Department. In the session on visioning, I adopted the following simple framework:

☆ forget about Finance as it currently exists in this organization;

☆ identify current assumptions around what Finance should be;

☆ don't simply project your assumptions of what Finance is into the future;

☆ consider the identified requirements for Finance in the organization; and,

☆ focus on an ideal situation where Finance can meet all the identified requirements, regardless of how possible this might be or how much time it might take to get there.

The resulting discussion was a lively one that probed many boundaries of what we all thought Finance was about. The resulting Vision Statement left Finance open to undertake a wide variety of initiatives within a scope defined by the group:

> The Finance group will be a pro-active business partner for the organization, providing actionable management information for decision-making. In addition Finance will be an active financial manager ensuring that effective financial processes and controls are in place to ensure appropriate financial accountability.

Like other Vision Statements, this one was somewhat lofty and vague. But in comparison to many other Finance Vision Statements that I have been involved in generating, this one had a more open and dynamic feeling. The TSK approach had given the group an effective framework for an open and free exchange of ideas not bounded by anything other than their own creativity, and the result seemed to me a step above the ordinary.

Conclusion

These personal experiments in TSK have produced interesting, effective, and above all practical business results. Increased awareness of your own positions, assumptions, and biases goes a very long way to improving your ability to accept change and to flow with it. Setting aside the specific experiments, it seems clear to me that a TSK investigation of perception and awareness in relation to time, space, and knowledge can provide both unique and helpful insights regarding real business issues and problems.

Some may see the use of the TSK vision in these business situations as trivializing the power of the vision, but I believe that such applications demonstrate both the power and flexibility of the TSK approach. TSK can deal with the most profound philosophical questions, but it can just as easily provide insight into basic business concerns.

A more serious concern could also be raised. Distilling the TSK vision into a consulting toolkit runs the risk of stripping it of its dynamic essence. The very power of TSK is its lack of prescribed framework and codified approach. I can see that this concern is real, but my own sense is that TSK is robust enough to accommodate both an applied approach and a pure theoretical vision.

TSK is a powerful vision that can be applied in a most practical manner. Its use of dynamic inquiry forms a solid foundation for creative investigation, whether the subject is personal perception or business processes. As Tarthang Tulku says:

> More important than any characterization [of TSK] is learning to make inquiry into a tool for understanding and a manifestation of knowledge. (DTS, p. xxi)

Through inquiry we can come to appreciate what Hamlet said (Act I, Scene 5, lines 166-67): "There are more things in heaven and earth, Horatio, than are dreamt of in your philosophy." Or, as I discovered in my own backyard, a tree is not just a tree.

TSK and the Ecology of Business

Ronald E. Purser

The fact that the United States is consuming over one-third of the world's fossil fuels while making up only five percent of the world's population is a stark symbol of our gross national imbalance. This pervasive sense of collective imbalance can be directly attributed to our obsession with security and our exclusive reliance on what *Love of Knowledge* calls technological knowledge.

On the brink of the twenty-first century, Western technological civilization is for the most part guided by a seventeenth-century Cartesian vision of the universe. The macro-economic structure in which modern corporations operate is informed by a mechanistic, linear world view. Within this context, the natural environment is 'objectified' and viewed as an externality that can be controlled, tamed and commodified for human consumption.Time within this view is also related to as a commodity, and is linearized in order to maximize market and production efficiencies.

As a result of this scheme, human beings within modern society have become entrained to the ever accelerating pace of technology, from the earlier repetitive rhythms of the steam engine to the lightning-quick speed of the microchip.

Ironically, while our ability to enclose space and control time in the natural world has increased exponentially since Francis Bacon's exclamation that "knowledge is power," it seems that Western civilization is lacking the knowledge which would allow us to live securely in space and time. Perhaps more important, while the eye of scientific empiricism has provided human beings with incredible power in controlling the forces of nature, Western civilization, in the words of Thomas Berry (1988), lacks a ". . . functional cosmology to guide and discipline our human use of all this knowledge and skill and all these energy resources." Degradation of the environment—as evidenced by global warming, acid rain, deforestation, air and water pollution, growing landfills, and the threat of nuclear war—is calling into question the reliability of technological knowledge.

As the instruments of technological knowledge become further refined and specialized, the isolated self finds itself confronted with a perpetual knowledge gap, always having to play catch-up. This is evidenced by the struggle to keep pace with the increasing rate of technological change, while our attempt to tame time's flow is observable in our obsession with efficiency and time-saving devices. Despite this headlong race toward the future, the frenzied pace of technological change seems to be guided by an impersonal and eerie autonomous momentum, as if a ghost in the machine were calling the shots.

Paradoxically, our attempts to control time and nature so that we can secure our status as autonomous and independent selves has intensified the momentum of time. Instead of feeling secure and at home in the world, we find ourselves resisting the changes that such an accelerated time brings. In order to remain 'in control' and maintain our identity or sense of permanence, we cling ever more tightly to our positions. We fortify them by increasing our preoccupation with production and consumption. Governed by rigid 'from-to' structures, the natural environment is relegated to an outside-stander, reduced to a bottomless pit and dumping ground for the industrial machine. Our cultural addiction to this stultifying pattern leaves little room for new knowledge to shine through. Based on observations of time-orientations of his clients in therapy, Herbert Rappaport (1990) concludes that most individuals, and even our culture as a whole, are lacking a viable life's vision.

The busy-ness of business and its insensitivity to the environment are symptomatic of a world view dominated by the rhythm of lower time and limited by the perspective of technological knowledge. Type-A behavior, a quick-fix managerial mentality, and the short time-horizons of corporate shareholders are indicative of entrancement with lower space, time, and knowledge. The concern in business with maximizing efficiency has alienated us from nature, the community, and our senses. Relying increasingly on models to guide our way, we find that our sense of common sense comes to seem alien. Direct knowing is antithetical to linearized efficiency.

In the modern corporation, models have become a substitute for direct knowing. They act as an intervening barrier between the environment, employees, and managerial awareness. Rather than trusting and valuing direct perception, we disseminate highly systematized (and in some cases, highly coercive) training techniques as models of the 'way to know'. Our attempts at control have actually cut us off from the fullness of space, the creativity of time, and the sensitivity of knowledge.

The greening of modern business institutions will require nothing short of a new functional cosmology and a more balanced embodiment of ourselves in space and time. The TSK vision offers such a liberating, functional cosmology, one that can awaken us from our cultural pathology by demythologizing technological knowledge. Our fulfillment is not to be found in our attempts to enclose space or control time, but in our 're-participation' in the intimacy of knowing space and time.

The perception of the environment as an 'economic resource', 'outside-stander', or 'thing out there' is maintained by a very specific focal setting on space.

Within this lower space perspective, knowing 'bottoms out'. It becomes increasingly preoccupied with objects, dimming our aesthetic awareness and intimacy with the natural world of appearance. From the theoretical perspective of polar knowledge, the environment is seen as merely a collection of distant objects.

The conventional knower separates himself or herself out from that which is known. Foregoing direct knowing, we take up a position and identify with the bystander-self, turning once more to models to describe the now distanced world. As our dependence on models proliferates, we no longer see, hear, smell, taste, or touch what is before us. With space reduced to a container for objects, even the air we breathe is taken for granted. As Tarthang Tulku states (LOK, p. 204): " . . . measuring the world in terms of self-concern, [the self] only sees what it has the tools to measure."

We are all very fragile creatures. The articulation of the concept of the biosphere has revealed the delicate interdependence of all life forms. Ecological consciousness arises as we open to a more inclusive space, dynamic time, and pervasive knowledge.

Working with the TSK exercises naturally restores our feelings of intimacy and reverence for the natural world. From a wider perspective we can see that even the earth and all its inhabitants are suspended in space.

The primacy of space is central to ecological consciousness. Within a more comprehensive space, knowingness reveals an open dimension to all appearance, thus pointing to the inherent insubstantiality of our consumption and grasping habits.

At the same time, by working with time, we can unfreeze our attempts to control nature for egoic purposes. We cannot stop the earth from turning on its axis or from revolving around the sun. While we may be able to partially tame time by erecting and maintaining rigid structures and fixed patterns (whether they be organizational or personal), such attempts at control are deadening and result in a loss of creativity. Time in its pure form is uncontrollable and closer to chaos. While embracing the spontaneity of time may at first seem threatening ('we' can't 'control' it), allowing ourselves to do so can inspire us to creative and authentic action.

By deepening our sensitivity and responsiveness to these spontaneous, all-pervasive energies, we can place our trust in knowingness and decrease our dependence on dualistic models. Rather than being entranced by the stories and narratives of the 'technological progress' myth, our guidance can come more directly from clarity itself. As our knowing is liberated from the confines and limitations of models, we can begin to see through them. A starting point is the realization that the very rational process that we exalt as the only true way to understanding is itself a mythic, imaginative dream experience (LOK, p. 205). Through Great Knowledge, we can come to see that "we *are* the tame, law-like rational seeming that is . . . magic." (TSK, p. 107, emphasis and ellipsis in original)

REFERENCES

Berry, Thomas, 1988. *The Dream of the Earth*. San Francisco: Sierra Club Books.

Rappaport, Herbert, 1990. *Marking Time*. New York: Simon & Schuster.

Inner Time Management

Steve Randall

Time is like a mirror—
it reflects different things,
depending on how we look at it.

The statement just quoted implies that time is not a constant, but somehow changes depending on our consciousness, and on our view of time. If this is true, perhaps we do not need to race against time, or see it as a scarce resource, in spite of what our culture teaches us. Perhaps the assumptions that conventional time management programs work with need to be investigated more closely.

Most conventional time management (CTM) workshops promise to help get more done with less stress. They generally rely on the following steps, all useful:

☆ Clarify and write down your long- and short-term objectives in major areas of life. Keep these lists of objectives current.

☆ Break projects down into doable tasks. Update project plans as necessary.

☆ For all identified tasks, set priorities and estimate the time required so that you are aware of what is important and when things are scheduled.

☆ Schedule periodically and create to-do lists and calendars with scheduled tasks and appointments.

☆ Do the tasks, focusing on top priorities, in the time allocated (except for unexpected changes).

☆ Periodically ask: "What is the best use of my time right now?" Change tasks as appropriate.

Some programs that center on these themes promise to lead to mastery of time. But how can we master time if we do not understand what it is? How can we get things done without feeling overwhelmed or anxious, if we do not know how time pressures us?

What Causes Time Pressure?

Time pressure is one of the greatest sources of stress for most people. Based on a survey he conducted with top and mid-level executives, Richard Winter reported (1983), "According to almost three of every four . . . executives . . . the most stressful situation was the one that occurred most frequently: work demands and time pressures." Referring to the general U.S. population,

Dr. Stephan Rechtschaffen (1986) wrote, "I would say that 95 percent of the stress in our lives relates to our feeling of time poverty." (Time poverty is the sense that you do not have enough time, that it is a scarce commodity.) And Larry Dossey (1982) wrote, "Many illnesses—perhaps most—may be caused either wholly or in part by our misperception of time. . . . I am convinced . . . we can destroy ourselves through the creation of illness by perceiving time in a linear, one-way flow."

What is the source of time pressure? It is easy to list candidates: demands from the boss, deadlines, a to-do list that is too full, disorganization and confusion about what to do, unclear priorities, lack of an effective scheduling system, and imbalance in the energies of the body. It would be useful for you to reflect for yourself on what you think the causes of time pressure are before going on, and see if you can add to this list.

All of these examples are in some sense external, even if they result from our own lack of planning or organization. But is that really the whole picture? For instance, consider deadlines. A deadline seems the most obvious cause of time pressure, starting with the deadlines we faced in school to finish assignments. We tend to think that pressure is somehow built into deadlines themselves, that there is something objectively 'real' about deadline pressure. But is it true that deadlines always cause pressure? Or is there some other factor at work? In my own experience, there have been many times when I felt pressured under a deadline and was able to turn the whole thing around, sometimes even ending up with a peak experience.

Have you had similar experiences? Please think about this a minute. Have you had a deadline where you felt significant pressure, then gotten involved in the work so much that the pressure changed and perhaps completely disappeared?

If you have ever had an experience like this, it suggests that the pressure really isn't built into the deadline. There is something we can do to change the experience. There is some other way to do it.

Pressure Is Related to Involvement

Staying with the example of a deadline, it seems that pressure somehow depends on the way we relate to the deadline. It depends on our perspective, on how much we are involved in what we are doing.

Let us take a little closer look at this idea. We typically use the word "involvement" very loosely. We have different verbs to indicate the degree to which we are involved in what we're doing. I might be 'holding back', which means I have something to do, but am resisting it. Perhaps I think about it, but then put it out of my mind. If I get a little more 'involved', I will 'resign myself' to doing the job. Basically that seems to mean I am not just putting it out of my mind any longer. With a bit more 'involvement', I 'get into' the job. A bit more than that, and I can say for the first time that I am 'really involved', meaning that I am 'tuned in' to the activity. Beyond that, I am 'absorbed' or 'engrossed'.

Now, here is a very important question to ask yourself. If you consider your past experiences, as you became more involved, what happened to the pressure? Please think about this a minute before going on.

Based on my own experience and the experience of many hundreds of people I've talked to, this "involvement scale" when applied to the experience of time leads to the following principle: The pressure we feel is directly proportional to how much we are resisting what we are trying to get done. If we are holding back or resisting, then there is pressure; if we relax, 'get into it', get really involved, there is no pressure. The pressure dissolves. Total involvement means no pressure; detached holding back means pressure.

Holding back is something like distanced observation, as in the old adage: "A watched pot never boils." If you take a point of view separate from things, distant from what is happening, time will seem to pass strongly. There will be an intense sense of friction and lack of control, no matter whether time goes quickly or slowly. On the other hand, if you do not take and hold a point of view, but get completely involved in whatever is happening, time will not seem to pass at all, or else will pass without friction or feeling out of control. This is when you will be most creative and productive.

This discussion of pressure leads directly to the heart of ITM, Inner Time Management. If pressure really depends on our perspective, there is something we can do about it. We can manage time internally; we can change the feeling of time passing by changing our attitudes and our approaches. We are not victims of time.

But before we look at this more closely, we need to ask some basic questions. What is time all about? At this point, I have a request to make. Please take some paper, set aside a few minutes, and address the question, "What is time?" Do some thinking about this and write some notes before going on. Even if you do not come to a definitive answer, it is very useful to begin seriously asking this question.

Types of Time

It can be useful to identify three types or 'faces' of time:

First, "event time" is the continual occurrence of physical and experiential events. The word "event" is used to describe something that happened or is happening 'now', like getting up in the morning, feeding a pet, a car honking its horn, or noticing that you are hungry. Event time is what we hear and see on the TV and radio news shows.

A second face of time is symbolized by tools for measuring event time. We measure length and distance in different ways—by means of rulers, yardsticks, and odometers in cars. And different cultures in the world measure event time in different ways. Most cultures use clocks and watches, and subtract a 'start time' from a 'stop time' to figure out 'how long' something takes. But there have also been cultures that use "the length of time it takes to cook rice" as a way to measure 'how long things take'. "Telling time" is knowing how to measure time in your culture, knowing how much measured "clock time" corresponds to the events that are happening 'now'. Most children's books on time are primarily about learning how to 'tell time' this way.

Since American-European cultures focus on measured time and events in physical time, time management in Western countries has become simply a matter of choosing, organizing, and scheduling events. The purpose of conventional time management (CTM) is to help us produce more in a given period of measured time, and thereby decrease some of the anxiety and pressure we feel about time.

But is that enough? Although time-management seminar graduates may be able to accomplish more as a result of their training, there is growing recognition that they still feel like they don't have enough time. Some even feel that things have gotten worse. As Stephen Covey (1994, p. 31) says, "Concerns about quality of life are just as likely to come from someone with a high level of [conventional] time-management training as from someone without it."

This suggests that instead of focusing just on events in time, on what we're doing, we would do well to also explore how things are going—the range of experience that runs from feeling overwhelmed and pressured on the one hand to things flowing so well we're not even aware of time passing. Exploring how it's going, the quality or feeling of time, is the domain of inner time management (ITM). It leads to the third face of time.

The third face of time is probably most important for our happiness, although it is also probably the face that is least understood and most undervalued. Here we will call it "felt time," though it might also be called experienced time. Felt time includes all the different ways we feel or experience time. We may feel time move quickly when we are having "a great time." During some of the best moments of our lives, things seem timeless, with little or no feeling of time passing. On the other hand, we feel time 'drag' or pass slowly when we are bored, or having 'a bad time'. And we feel anxious about time when it seems we do not have enough of it.

Optimizing Feelings of Time

This is where ITM comes in. Rather than the usual CTM focus on what we want to do, ITM gives methods to optimize the moment-by-moment feelings of time, including the way we relate to our current activity. ITM teaches how to increase involvement by moving through the stages of involvement that were already introduced above:

(1) holding back from doing something;
(2) resigning ourselves to doing something;
(3) getting into it;
(4) being involved;
(5) being engrossed or absorbed.

By moving toward the complete absorption of peak performance, we can transform troublesome feelings of time passing—including 'overwhelm', time pressure, frustration, boredom, and anxiety about not having enough time.

Learning and consistently using both CTM and ITM methods is valuable for almost everyone. CTM can help us get more done, but ITM helps us improve the quality of our lives. Neither CTM nor ITM by itself resolves all our issues with time. But by combining the discipline of planning and organizing what we do with methods of improving the way we do things, there is no limit to our productivity and well-being.

"Western Standard Time" (Linear Time)

The usual way that most adults in the West experience time is as linear time. This is an experiential perspective completely independent of measured time. It combines (1) the actual feeling of time passing in a linear and directed way from one moment to another, (2) the separation of past from present and present from future, and (3) a sense that you are positioned only now, in the present.

In the linear view, time is like a conveyor belt that moves horizontally at a constant and unchangeable speed between past, present, and future 'rooms' in our experience. These different aspects of time are actually felt to be separate. We are located in the present 'room'.

Linear time is a major feature of our Western cultural world-view. It may have originated with Newton, some three hundred years ago, since Newton portrayed time as an absolute physical reality, and regarded the passage of time as independent of consciousness. From a linear-time world-view, it makes no difference what you think, feel, or do, or how you look at time—time does not change. As a result, we may feel somewhat helpless in the face of time. It seems our only choice is to adapt to this temporal 'reality', finding ways to work 'within' its limitations as skillfully as possible. This is the basic model that CTM programs work with.

Despite this 'standard' understanding that time is independent of consciousness, we somehow also feel a flow of time within ourselves, an internal mirroring of the 'real' external flow. This is considered the norm. In fact, we use the phrase "losing track of time" to indicate a kind of negligence when our internal feeling of time passing (FTP) doesn't accurately 'track' the presumed external flow of time. Our sense of time (SOT), the ability to estimate what the clock time is, seems connected to our FTP. It is often believed to result from an internal biological mechanism that tracks or measures external time flow.

It is the FTP that sets up the host of familiar problems mentioned above, relating to time pressure, anxiety, time poverty, etc. But as long as the FTP is thought to simply mirror external events, it is considered to be unchangeable, like the presumed constant external flow of time. So we are left with the prospect of—at best—adapting to an uncontrollable flow of time and its attendant pressure and transitoriness. Is that really the only choice we have?

A Case Study: Procrastination

Procrastination offers an interesting example of how our feeling of time passing is created and strengthened.

Suppose I have a speech I want to prepare. It is Monday, and the speech is scheduled for Thursday. I have a four-hour block of time today that I can use to prepare the speech, and no other open time before Thursday. Now is definitely the best time to work on it if I want to do a good job.

I begin working on the script for the talk. The work goes reasonably well. Sentences flow; the writing goes almost by itself, effortlessly. Before long, I am so engrossed in the task that I am not aware of any feeling of time passing. Nor am I aware of past, present, or future. There is only timeless absorption in my work.

Eventually I get a little confused about the message I want to communicate. Because I do not face the confusion directly, my mind starts to wander. I look at the clock and realize it is almost time for my favorite TV show. Pretty soon I am thinking about how I might be able to finish my preparation right after the show is over, before I go to bed. Yes, it seems possible! I think I have enough time. So I put my work aside and settle down to watch the show.

But once I have started to procrastinate, the quality of my experience suffers. Watching the show is not as enjoyable as I had hoped it would be, because awareness is divided between watching the show and being aware that I have to do my work. Time is passing relentlessly, and it feels like the future is closing in on me. I am watching TV here in the present, feeling anxious and guilty about a job waiting for me in the future. In addition, I have missed an opportunity, and feel less confident and capable as a result.

So as a result of procrastination, I end up with a stronger FTP, together with a greater separation between present and future. When I started my work, I really concentrated. The experience was timeless. Now, after procrastinating, my sense of time has changed for the worse. If I take the standard view that time is simply a physical reality, flowing at a constant speed independent of our consciousness, I can fool myself into thinking that procrastination is simply rescheduling a task to a different container on the conveyor-belt of time. But that is not the truth of the experience. My procrastinating has actually created the conveyor with its containers, along with the feeling of being out of control of time's flow.

This deeper analysis warrants redefining procrastination as follows: Procrastination is the repression or suppression of an unpleasant feeling that results in temporally separating oneself from a task. We are familiar with spatial separations, where we build walls or just walk away from something. But we can also temporally separate ourselves from things that we do not like. What seems to happen is that the energy of the feeling we dislike (confusion, in my example) is pushed away and transformed into a felt experience of time passing. The energy is not lost, it is just changed to a different form.

Rather than measuring or mirroring a presumed 'external flow' (which scientists haven't found), our FTP seems to measure how much we are resisting what we are doing. It is the aggregate result of resisting past negative experiences. In this sense, the commonly perceived structure of time is actually a transformation of energy we dislike. We could even say that repressed energy is all that constitutes our common experience of time. To put it differently, our sense of time passing is not an accurate mirroring of external forces or events; instead, it measures our separation from whatever we are doing. When our FTP grows out of past resistance, no wonder our typical experience of time is discomfort and frustration.

What this means is that full schedules and modern speedy technology are not the problem. This is a good thing. We can learn to dismantle the FTP, gradually moving toward the timelessness and fulfillment of peak experiences. But to do that, we have to question linear time.

The Construction of Time

The experience of linear time is very common in our culture; however, felt time is different in other cultures. Anthropologist Edward Hall (1983, p. 4) says, "Each culture has its own time frames in which the patterns are unique." For instance, the Navajo consider the future uncertain and unreal, so a Navajo is often neither interested in nor motivated by 'future' rewards (p. 28). Similarly, Hopis often have no timetables or schedules in their minds for ordinary projects, such as building a house. Even in European cultures there are differences: Northern Europeans tend to do one thing at a time in accord with a fixed schedule, while Mediterranean cultures stress involvement of people and completion of transactions rather than adherence to preset schedules, and appointments are not taken all that seriously. In comparing East and West, Hall makes a more general point (p. 91): "In the West . . . time is an outside force helping us to organize our lives. In the East, time springs from the self and is not imposed."

Felt time depends on the culture, but it also depends on individual learning within a culture. As infants we have no sense of time flowing in a linear way. The sense of linear time is developed gradually, and child-development studies suggest that it is not fully present until about the age of seven.

How does linear time get set up? Drawing on the descriptions found in Tarthang Tulku's *Kum Nye Relaxation* and TSK books, here is a quick summary: A feeling begins to arise in awareness. But rather than *feel* the feeling, we turn away from it. The feeling is repressed or suppressed, and we lose a measure of confidence as well as a bit of the natural fulfillment that accompanies being fully involved in our energies. The energy flowing in the heart is reduced, and we feel somewhat pressured. Excess energy flows to the head. As our thinking skips here and there in the separate past, present, and future rooms of our experience, a sense of detached self-consciousness intensifies. Energy in the throat, which is closely associated with time, becomes agitated as we become anxious and more aware of time passing. We feel more helpless; time becomes more threatening, a greater enemy. There is a sense of a dissatisfied self confined to its one small moment of time, reaching out to seek satisfaction through various objects and activities or looking to other people to fulfill desires. The self 'looks forward' to things but has difficulty fully appreciating them. We always have to move on.

The linear view of time is perhaps the least productive and healthy way to experience time. It usually involves anxiety, pressure, and friction, together with a division of awareness between the present job and concern with the future. There is always wasted mental and emotional energy.

But the linear experience of time is not the only option available to us. Instead of linear time, more enlightened individuals apparently somehow learn to find a sense of stillness within all activities. For example, Tarthang Tulku (TSK, p. 136) writes, "Time is neither linear nor sequential; in fact, there are neither moments nor successive movement, and thus no succession." Can we arrive at an understanding or experience closer to this description? Can we dismantle the FTP, gradually moving toward the timelessness and fulfillment of peak experiences?

The Optimal Way of Experiencing Time

Is there an optimal way of experiencing time? For an answer, you can check your own experience. What was your experience of time during peak experiences, the best moments or periods of your life? During peak experiences of love or meditation experiences; during peak productivity; or in your best sports performances? Learning to rely on your own experience is important, so think about this question for a couple of minutes before reading on.

I have asked thousands of people the question in the previous paragraph. Again and again they say something close to this: "In those situations, time doesn't flow, at least not in the 'normal' way. I'm either not aware of time passing at all, or it seems to go fast, but without a feeling of being out of control."

This is very interesting. During their best moments, people are not feeling time in the 'standard' way, where time's passing is out of our control, making us feel anxious. In this kind of experience, things are happening ('event time'), yet 'felt time' doesn't have an arrow—it doesn't have a sense of flowing forward into the future.

If you think about it, this is a common experience. The things that we really enjoy and get involved in are 'over before we know it'. This is a very particular kind of felt time, often called 'timelessness'. When we're completely absorbed in something, totally engaged or involved, there is a sense of being very present with what is at hand, and time doesn't seem to pass in the 'standard' way. Timelessness is a kind of 'felt time' that actually doesn't seem to 'pass' at all.

Abraham Maslow noted this phenomenon in his best known book (1962) "[There is] the frequent report, especially by lovers, of the complete loss of extension in time. It is as if they had, in a way, some place in another world in which time simultaneously stood still and moved with great rapidity." As described by Linda Ackerman (1998, pp. 125–26), for the peak performer there is "a unified flowing from one moment to the next, in which he is in control of his actions, and in which there is little distinction between self and environment, between stimulus and response, or between past, present, and future."

Clock-Watching: An Experiment in Timelessness

Timelessness can be a fleeting experience, easily missed. Focused in certain directions on goals that we have set, we are seldom very aware of the more subtle qualities of experience, especially the changes in the nuances of time passing.

Here is an exercise that provides a stable environment, so that you easily see variations in 'felt time'—the changing feelings we have with, within, and about time. It may also clarify the differences between measured, or clock time, and felt time. The exercise balances energy in the throat, where imbalance seems to produce pressure and anxiety about time (see TSK, ex. 24), as well as the left and right hemispheres of the brain (as shown by research in applied kinesiology).

Set up your environment so that you have a quiet place and at least ten minutes when you will not be interrupted or distracted. Allow five minutes for the exercise and five minutes afterwards for reflecting and writing. Set up a clock to watch, preferably a large clock with a second hand that is seven to eight feet away from where you will sit. Watch the second hand, concentrating loosely as you watch the hand move. Breathe easily, gently, and smoothly through both nose and mouth, with the tip of your tongue on the upper palate just in back of your front teeth.

Watch for five minutes or so. It can be helpful to set a timer for five minutes, since that will relieve you of the need to track clock time. As you continue, see if you can let the breath become more and more even and continuous, without breaks or jerkiness—this is important. Evenness and continuity of the breath correspond closely to the clarity and peacefulness of awareness.

As you relax and observe the movement of the second hand, you might explore in a relaxed way these questions: Does the FTP change? If so, how? If there were any feelings of pressure and anxiety present at the outset, what happened to them? Did any new feelings of pressure or anxiety arise? Any other feelings? How is the quality of breathing related to flow of time? How is the feeling of a separate self or observer related to the feelings of time passing?

At the end of the five minutes, make some notes about what happened—anything that strikes you as interesting or worth writing down. Did time feel out of control? Any anxiety about time? Did the speed of time change? If so, what changed: event time as measured by the clock, or felt time? Was every minute equally long? If you felt more relaxed at the end of the five minutes than you did at the beginning, reflect on what this means. If you can relax while watching the clock, you can probably stay relaxed in nearly any situation!

By balancing our breath and attention, we change our perception of time, lose awareness of time passing, and transform anxiety and pressure about time and deadlines. The type of breathing used in this exercise (borrowed from T. Tulku, 1978) is the most effective means of working with anxiety or pressure that I have found. You can practice it while sitting still, but you can also do it while involved in any activity.

What Inhibits Timelessness?

Having felt the anxious and pressured experience of linear time, and having occasionally known the freeing timelessness of peak experiences, we might very well ask, "What is it that keeps us from having more peak time-experiences?" This question can be answered in numerous ways. Here are three.

1. Western cultures implicitly teach that linear time is 'normal'. They also confuse event time, measured time, and felt time, providing only one word for time. And they implicitly teach that we should always feel time flowing constantly (or we're 'losing track' of time). This makes it difficult to facilitate peak experience, which has a timeless quality, or to allow for a sense of time passing that is free from friction.

2. Techniques to change 'felt time' are not common knowledge. Numerous meditation techniques evoke a level of timelessness, though they often simply try to ignore time's passing rather than examine exactly what time is. But apart from that, very few people teach methods to optimize our felt time. This is an area

where a positive contribution can be made. The simple breathing practice described above is a good example.

3. We avoid 'negative' feelings and sensations. Western cultures implicitly teach that turning away from or even suppressing 'negative' experience is normal and natural. However, turning away from 'negative' experiences strengthens the linear-time experience of time passing. Repressing or suppressing the energy of such feeling produces a stronger sense of time passing, whether it flows more slowly or more quickly. It helps to recognize that relating to experiences as 'negative' or 'positive' (where 'positive' is understood as the opposite of 'negative') is only one of numerous possibilities. For instance, one way to engage what is happening is to try to find, and be at, the center of 'negative' feeling, rather than observing (or avoiding) it from a separate position. Tarthang Tulku (1978) presents various exercises to explore this approach. (One that I have found especially effective is called "Heart Gold Thread.") The basic principle here is one I have already referred to: If you take a point of view separate from things and feel distant from what is happening, time will seem to pass with a strong feeling of friction and lack of control. On the other hand, if you get completely involved in what is happening, time won't seem to pass at all, or will pass without friction or feeling out of control.

Levels of Mastery of Time

We've taken a look at linear time, the 'normal' way of experiencing time in Western countries, and we have discussed optimal ways of experiencing time. But there's a wide range of possible ways of relating to time—many different levels of mastery of time and time stress. Here are six levels I have identified, with a short description of each:

1. *Struggling Against Time* Time is outside us, and we race and struggle against it. We are victims of pressure, overwhelm, and anxiety, and we think that this is normal and cannot be changed.

2. *Wondering About Ways to Relate to Time* Our relationship to time has loosened up, and we're wondering about the possibilities. We no longer feel consistently pressured and anxious.

3. *Seeing Time As An Ally, Not an Opposing Force* Beginning to see how our experience of time is created, we are able to apply various methods to transform some time pressure and anxiety.

4. *Allying with Time* By seeing different levels of time, it becomes possible to reduce time stress by half, at least when you can take a break or relax.

5. *Empowering Time* You are aware of subtle pointings, including a separating of self and other. Now it is possible to reduce stress by half even in the midst of stressful situations.

6. *Abiding in Time* At this level, you never let time stress get established. You abide in the peaceful yet energetic eye of the whirlwind of activities.

This introductory presentation might give you some sense of at least the first few stages in this progression.

But to go further, you will have to do a lot of questioning and investigation. It is important to do the exercises I have mentioned, or other exercises in the TSK books that loosen up the experience of time. Otherwise, it will probably take much, much longer to develop the necessary insights and make the required changes in habitual ways of seeing and experiencing time.

Embody a new relationship to time and your experience can shift dramatically. As Stephan Rechtschaffen (1996, p. 20) says, "Understanding time with your brain isn't enough . . . . When you learn to embody time, when you can shift it at will, then you will experience a wholeness, a freedom—time freedom . . . . ."

Two Closing Questions

Birds need two wings to fly. One is not more important than the other; they're simply both necessary. Similarly, it seems that to measure progress in life, we need to periodically consider two questions. One question is, "Am I doing the right thing?" A second one is, "Am I doing things right?"

Another way of stating the first question, mentioned early in this article, comes from time-management guru Alan Lakein (1973, p. 96): "What is the best use of my time right now?" This is a CTM question. Perhaps we haven't really thought much about what we're doing, and now it would be better to do something else, to take a different approach. It seems that we are often thinking implicitly about Lakein's question anyway—why not make it explicit?

The second question can be be rephrased in this way: "Am I timelessly involved in what I am doing?" This is an ITM question. It invites you to compare your experience to what people report about peak experiences of all kinds—that there is no sense of time flowing in a way that feels out of control. Ask this question periodically and reflect honestly on the result. If you are not totally involved, or if you feel time passing in a way that has even a slight bit of pressure or anxiety, there is room for improvement in both productivity and well-being. Look back at the exercises introduced here; perhaps they can help you identify what is between you and a timeless way of being and doing.

By finding the peaceful, yet most productive 'zone' at the center of our whirlwind of activities, we can transform our feelings of time flowing. We can let go of pressure, anxiety, and boredom, and we can get results "in no time." Why not take the chance? Why not invest the time?

REFERENCES

Ackerman, Linda S., 1998 "The Flow State: A New View of Organizations and Managing," in J. Adams, ed., *Transforming Work*, 2d. ed. Alexandria, VA: Miles River.

Covey, Stephen R., 1994. *First Things First.* New York: Simon & Schuster.

Dossey, Larry, 1982. *Space, Time and Medicine.* Boston & London: Shambhala.

Hall, Edward T., 1983. *The Dance of Life*. New York: Doubleday.

Lakein, Alan, 1973. *How to Get Control of Your Time and Your Life*. New York: Signet.

Maslow, Abraham, 1962. *Toward a Psychology of Being*. Princeton, N. J.: Van Nostrand.

Rechtschaffen, Stephan, 1996. *Time Shifting*. New York: Doubleday.

Tarthang Tulku, 1978. *Kum Nye Relaxation*, Parts 1 and 2 (Berkeley, CA: Dharma Publishing).

Winter, Richard E., 1983. *Coping with Executive Stress*. New York: McGraw-Hill.

EXTENSIONS

In the brief article that opens this section, Arnaud Pozin describes the TSK vision as offering a special kind of knowledge that is neutral with respect to all positions, a knowledge well attuned to the 'planetarization' of human culture and society in our time. This seems a fruitful insight, one that could serve to introduce this entire volume. It leads nicely into the three remaining articles, which explore the relation of TSK ways of knowing to the political realm, causal ways of thinking, and the methods and approaches of physics, regarded as the exemplar for all the sciences.

Like Pozin, Steve McCarl is interested in the political significance of a way of knowledge not bound to the self. He explores the implications via a comparison with Michael Oakeshott's suggestion that humanity speaks with different voices, and that it is important to bring all these voices together in a great conversation. Could TSK add a new voice to the mix? Could human beings—speaking without regard to the interests of a self, but with due regard for the importance of friendship—open a new kind of dialog in the political arena? These are the kinds of issues that this article brings to the foreground.

Ingrid Harris is concerned with an issue in some respects still more basic: the structure of causality.

Drawing on the classic presentation by David Hume, which challenged the 'reality' of the causal principle, she investigates the claim of scientific materialism to have worked out the logic of causality in ways that exclude all other options. She invites the possibility of causal *experience,* and finds in such experience the possibility of new ways of relating to time and space–new ways, that is, of being free.

The final article in this section, by Maxim Osinovsky, considers in some detail the relation between the TSK vision and scientific ways of understanding. The author points out some ways in which current understandings of space and time are compatible with the TSK vision. He suggests as well that TSK potentially offers a way out of the impasse in which science finds itself when it comes to making philosophical and existential sense out of its own findings, and calls attention in particular to the holistic understandings associated with such thinkers as David Bohm. In the end, however, he concludes that TSK is too impressionistic in its approach to attract much interest among this generation of scientists, and invites an active exploration into new forms of integration.

TSK and the Neutrality of Knowledge

Arnaud Pozin

Will we be able to find the knowledge we need to live together in peace? Habitual answers to this question emphasize a certain image (description) of what reality should—or could—be if we were to behave as human beings. And a lot of energy is used to 'make' reality conform to that description. In proceeding this way, many problems arise, among them the fact first that we will never completely succeed in reducing reality to one simple description; and second that the holders of different descriptions then fight with one another.

TSK offers a sound contribution in proposing, not a new description of reality, but an instruction (knowledge) concerning how we relate to reality. If we were to take the example of driving, we could say TSK does not describe where we are supposed to go, but proposes instructions about how to drive, each 'destination' being the untouchable freedom of 'each' being.

Descriptions are given as long as it is believed that people are blind and separated from reality. But isn't such a belief the very definition of ignorance and the sign that instructions are needed? We could say that TSK is not preoccupied with what should be, but with what is and how to appreciate 'it'. In this sense, instructions are given when there is a lack of appreciation. Or, to put it in other terms: knowledge 'comes' when/where knowledge 'is lacking'.

Time, Space, and Knowledge are neutral 'concepts'. They offer the possibility of describing reality in a way that fully respects but is not bound either to religious faith or scientific doubt, a way that itself is neither religion nor science. We are close here to the Occidental realm of neutrality: the freedom to study all religion, all science, all people, all medicine, all political or economic systems. Here is the power of neutrality co-emergent with the freedom to know.

This freedom to know is a 'federalist' concept of crucial importance for our time. It offers a way out of every kind of false integration. We could say the survival of mankind and the successful negotiation of the process of 'planetarization' are closely related to and depend on our ability to preserve and let grow new, neutral ways of knowing that respect ancient collective identities, but are rooted in our new planetary identity. We need more accommodating and neutral 'concepts' to counteract the auto-dissolution produced by the 'never-ending' fight between both personal and collective identities.

There is a neutral way to speak—and relate to reality—that does not harm any identity, and in fact unites all identities in the collective hope for peace. This possibility—both outwardly and inwardly, personally and collectively—is a gift of Great Knowledge (here understood as the basic knowledge common to all beings). Neutrality is in fact nothing more than an 'expression' of emptiness. When 'our' mind is empty of underlying expectations, we find the power (time) of neutrality. Since we all share the basic right to be, it is more than enough to find a neutral common ground from which the wisdom of our diversity may grow.

Neutrality is often understood politically as the power not to interfere, but here this is not the intention. Neutrality is rather a way to be and act without a grasping attitude: an art of expressing and respecting what we have in common, amd a way to find together a common source from which flows the strength to share. It is an encompassing way of being human, an aptitude to emphasize what connects us rather than what divides us, and perhaps, if we are allowed to dream on Planet Earth, a chance to reach together the love we need.

TSK AND THE PUBLIC

Steven R. McCarl

A fundamental issue for political theorists, given the importance of the individual in the practice of politics as well as in modern political thought, is the nature of individualism. Particularly important is the skill and willingness individuals display for forming unifying relationships.

When individuals through a process of interaction and relationship become a 'we', unifying relationships are set in place. In such circumstances, a 'we' may be a public, an association that makes possible accomplishments requiring cooperation and coordination on a relatively large scale. Under what circumstances does this happen? What must we examine to understand the formation of a public 'we'?

There are many ways of posing this issue. Hannah Pitkin (1981, in Petranker, 1997, p. 235) asks, "How shall we understand ourselves as simultaneously both private and public beings?" Tracy Strong (1992, p. 4) seeks "to explain in what manner and by what processes a particular understanding of the self (the person, the agent) is organized as, implicated in, and legitimizes a particular vision of the political order." Strong notes that one could just as well move in the opposite direction—from the political order to the self—as the two are interdependent.

In modern liberal states, the basis for the political seems to be not a sense of 'we', a common good, but rather a "free contract among rational individuals for the protection of their natural rights" (Smith, 1999, p. 625). Natural rights entitle the individual to pursue his/her self-interest with a minimal sense of a 'we'. The public is hardly a public in the sense of a people joined through a set of strong unifying relationships. In particular, the public in the liberal state is intentionally neutral with regard to the nature of the common good, so that individuals can pursue their private individual interests with minimal public or private interference.[1]

The liberal model allows for significantly different norms in the public and private spheres. Evan Charney (1998, p. 109) points out that in liberal states a person may abide in private life by the rules of a hierarchical church authority, yet in public life participate in (and be subject to) an egalitarian open process of collective will-formation. Within liberal theory the public/private distinction is deemed necessary to respect difference and support pluralism. It contrasts strongly with the view that public and private must be one, as espoused, for example, in theocratic regimes and sects.

As Tracy Strong's comments above indicate, there is a necessary relationship between the nature of the individual and the nature of the associations the individual forms. As students of TSK will be aware, the private self is a restricted sort of being, and a public based on such private selves will be a restricted public. In this paper I explore the nature of a public comprised of individuals reoriented by the TSK vision toward a less private way of being—a public that would differ significantly from the model provided by liberal theory. In this connection, I develop the notion of the "transparent I." I relate this understanding to Michael Oakeshott's views on "The voice of poetry in the conversation of mankind," reflect on how the notion of the transparent I can illuminate the TSK vision, and include some thoughts on how TSK converges with the natural law tradition.

The Self and the Conversation of Mankind

Michael Oakeshott describes the self as primordial activity, there being no "condition of rest or passivity to be overcome before it can begin" (1991, p. 496). The self is what is noticing, or imagining. Since without it nothing appears in consciousness, from the very beginning the self seems to be unavoidably present.

Tarthang Tulku also points to this primordial quality and connects it to the self's self-perpetuating nature. He notes that the connections within a narrative are provided by the self: I felt this; such and such happened to me; they did such and such, I noticed; I experienced x, y, and z, etc. Tulku points out that where the self is present, "All knowing is directed toward points after the flow of self-centered experience is already in operation . . . ." (LOK, p. 174).

Inherent in the self is its demarcation of itself from all that is "not self" (Oakeshott, 1991, p. 495). Robert Nozick (1989, p. 144–47) concurs: he describes the self as a bit of consciousness that is unlike all other bits of consciousness (except other selves) in being aware both of other bits of consciousness and of itself as a location of consciousness (i.e., as reflexive). The self recognizes its own activities not merely as the activities that are generally those of selves, but as the activities of a unique self: "it's self." The self says, "This is mine," and "Such and such happened to me." It appropriates, acquires, and identifies itself with various experiences and images. It is as though the self has sovereignty as it establishes the boundary between itself and not self.

The self, according to Oakeshott, appears in a specific mode of action correlated to the kind of images to which it is currently related. Its nature varies as the classes of images it is connected to vary. The self "is constituted in the activity of making and moving among images." The not-self to which the self is always in relationship "is merely a certain sort of image recognized as such because it behaves in a certain manner and responds to our questioning appropriately" (1991, pp. 496-97).

Oakeshott distinguishes three major classes of images to which selves are connected. He calls these three "voices," or modes of imagining; together they comprise "the conversation of mankind." The three are the scientific, poetic, and practical.[2]

In principle the three voices do not intermingle. Images cannot appear in any universe of discourse other than the one in which they do; "that is, each mode begins and ends wholly within itself." 'Water', for example, is an image within the discourse of practicality. H_2O, in contrast, occurs within the discourse of science. The rules governing the behavior of H_2O are not the same as the rules pertaining to 'water'. A scientist does not resolve 'water' into H_2O; H_2O is of a wholly different discourse. To speak of H_2O as the chemical formula for water is confusing, for it assumes that the two modes of discourse can intermingle, that the images of each are interchangeable, and that they have the same fundamental nature (Oakeshott, 1991, p. 514). The meaning of 'H_2O' requires knowledge of an elaborate and complex rational context: science. The meaning of 'water' does not.

The voice of practical activity is by far the most common current idiom, nearly ubiquitous for many people. Thus, the most common kind of self is the practical self. Hobbes was perhaps the first to recognize and describe the practical self in a systematic manner. He captures a basic quality in this comment: "Felicity is a continual progress of the desire from one object to another, the attaining of the former being still but the way to the latter" (1958, p. 86). The self and never-ending desires are inseparable. The constituents of the world of practicality are images of pleasure and pain; the habits of the self are habits of desire and aversion.

In the practical world, images are factual if by regarding them as fact "the desiring self is preserved for further activity" (1991, p. 499). This aspect of practicality is closely related to Harvey Mansfield's point (1995, p. 56) that interest is unerotic. It is not so much that images (the not-self) are attractive as that the self's motives push the self toward objects. The self remains 'in control' as it methodically goes about satisfying its desires. Erotic love, in contrast, is an irrational prejudice in favor of another that makes one forget oneself—not a likely development for the practical self.

The self "occupies a world of its own, a world of images related to its own desires" (Oakeshott, 1991, p. 500). That is, much as Hobbes suggested, the very nature of the self's world of practicality entails solitariness. The preoccupation of the practical self with fulfilling its desires leads the self to recognize the subjectivity of other selves, because if it does not it is less likely to realize its own desires. However, the self is predisposed to see other selves as not-self rather than recognizing them as selves who are also sources of subjectivity and imagining. Thus Oakeshott remarks, again in a Hobbesian manner, that ". . . desiring selves enter into no obligation, recognize no right; they admit the subjectivity of other selves only in order to make use of it for their own ends" (1991,p. 501; compare Nozick, 1989, p. 145).

Given the practical self, the public, understood as the common authority of the state, will be severely limited. Mansfield describes the limited role of the Hobbesian or Lockean state within the liberal doctrine of interests:

"It does not try to guide [interests] or perfect them, and so it is not allowed to connect or organize people's interests, thereby challenging their separate plurality. Each must be free to be flexible, now following one interest, now another." (1995, p. 57) In Oakeshottian terms (1991, p. 380), governing amounts to "seeking the most practicable adjustments for the collisions of individuals . . . ."

Practical discourse makes use of words as symbols, which are meant to be fixed in their meaning. It "is like a coinage, [and] the more fixed and invariable the value of its components, the more useful it is as a medium of exchange" (1991, p. 503). Here, however, the distinction between the practical voice and the voice of science starts to emerge. The world of science also contains symbols, but they are like chessmen: "They stand for what can be done with them according to known rules, and the rules which govern the use of mathematical symbols are more strict that those which govern the use of words." The rules of science are designed to exclude the private, esoteric, and ambiguous. The discourse of science is conditioned by the impulse to "construct a rational world of consequentially related conceptual images. . ." (1991, p. 508). This is a far different impulse than that of the desiring self, which operates in the discourse of practicality .

The third kind of discourse is that of poetry. By 'poetry' Oakeshott means the activity of making images and moving about among them in a contemplative or appreciative manner, with no regard whatsoever for their being factual or non-factual, worthy of approval or disapproval, or useful or useless. We care not, Oakeshott notes, if Anna Karenina said the words Tolstoy has her say; if David was of the proportions Donatello gave him; if Hamlet went to bed at a certain time; and so on (1991, p. 519). The poet only delights in making the images she does.

A poetic image is not intended to evoke a certain experience in others, for that would be attempting to satisfy a desire of the self, a concern of the practical realm (p. 523). Nor is argument or inference involved. "Since there is no problem to be solved, no hypothesis to be explored, no restlessness to be overcome, no desire to be satisfied, or approval to be won, there is no 'This, therefore That,' no passage from image to image in which each movement is a step in an elucidation or in the execution of a project" (p. 513). The playful character of the poetic idiom springs from "the self-sufficiency enjoyed by each engagement . . . and . . . the absence of any premeditated end." Poetic images are not read as symbols (governed by certain rules and having agreed-upon uses), but "are made, remade, observed, turned about, played with, meditated upon, and delighted in." (p. 517). "It is only when feelings are imaginary (in the sense of not being felt) that they become the stuff of a poetic image" (p. 524). In short poetic imagining seems to occur when the practical or scientific self is temporarily asleep or inactive—when "perception is unclouded by the preoccupations of desire, approval, curiosity, or inquiry" (p. 522–23).

Taken together these three voices comprise "the conversation of mankind." The nature of each voice is illuminated by comparison to the other two, but Oakeshott maintains that the voices cannot be arranged in hierarchical order. He warns against any voice becoming dominant, self-absorbed, and unappreciative of the other voices. The final measure of intellectual achievement is not within the parameters of any particular voice (especially not the practical), but rather in the contribution to the meeting place of all the voices: the conversation itself. This places a premium on the ability to enter fully into, and exit fully from, each voice. To the extent we are stuck in one idiom we can neither appreciate the whole conversation nor understand each of the composite voices (pp. 489–90).

I will argue below that the TSK vision, understood in light of the concept of 'the transparent I', constitutes a fourth voice, an extension of the poetic voice. This fourth voice provides for an appreciation of the whole conversation, as Oakeshott suggests. At the same time, it forms a basis for a model of the public radically different from the model liberal theory provides.

TSK, the Self and the Transparent I

Recall that a fundamental characteristic of the self is its separation from the reality (the not-self) it knows. The self projects its knowing capacity out into the world of the not-self. Knowledge is not out there in the not-self reality; it comes from the self. Thus the self becomes the owner, perceiver, and narrator of knowledge and experience—a bystander to the not-self reality (LOK, p. 264; Petranker, 1997, p. 245). This knowledge the self possesses of the not-self occurs in the form of substantive content—e.g., "x is stable;" "so and so is devious;" "it is cold outside"; "the bill will pass;" etc.

Is there an alternative to the self as knower? If knowledge were not what the knower/self knows (LOK), if individual agents did not have interests or positions, could they speak with a different voice? If so, could a different public emerge, one that would be more than a reflection or correlate of the private self?

To approach the possibility for a different kind of self, it is helpful to make a distinction between thinking as the process whereby thoughts or images appear and the thoughts or images as the content or end-product of thinking. Naming, identifying, abstracting, and generalizing are all ways of thinking that result in substantive content. Jack Petranker (1997, p. 264–66) points out that, "When thinking takes hold of experience and 'makes sense' of it, it turns that experience into the content of thoughts."[3] The self first turns experience into the content of thoughts, then becomes fixated or preoccupied with such content. This is consistent with Tarthang Tulku's observation that "in a single moment, the self posits a situation, takes up a position, and imposes meaning" (LOK, p. 147). The self constitutes itself as a position only by identifying with content (e.g., I am such and such; I have such and such; this property is 'mine'; etc.) Content makes a position possible; without it the self would have nothing with which to identify.

Tarthang Tulku explains (LOK, p. 195): "In knowing, 'I' observe, categorize, and characterize; 'I' apply names and labels. This activity seems to be what 'I' label as 'I'" (cf. Petranker 1997, p. 269). The self establishes its existence by generating content—in Oakeshott's terms, images—and then acts as if these images were completely independent of it. The images seem to take on a life of their own.

Recall Oakeshott's observation that fact and not-fact are practical images by which the self perpetuates itself (1991, p. 499). Facts are images of not-self that confirm the existence of the self to the self. Without the not-self there is no self. To distinguish itself from the not-self, the self must take a position, a position that is distinct from other positions (or "bits of consciousness," in Nozick's terms): "I am this, not that." This act of identification entails the availability of content and identification with content, or rather, with some bits of it.

The TSK vision provides an alternative that relates back to the process of thinking rather than to the content of thoughts. The focus is on the arising or sheer presence of phenomena rather than on the phenomena as content. In this view, the self becomes the space or place where knowledge appears, and knowledge becomes the arising of content rather than solely the content itself. As the agent of this knowledge, the self does not so much possess knowledge as express knowing. When knowledge is not what the knower knows, the knower becomes an instant of knowledge occurring, rather than one who knows such and such content. The continuous flow of ongoing knowing is aborted when knowing identifies with content. In Oakeshottian terms, the self is a maker of the world of images within which it exists, but upon identifying itself with images within one set of images (e.g., practical, scientific, poetic) it loses its power to make and appreciate images on an infinite scale. It thereby loses touch with the whole of reality; it becomes estranged, closed down, and narrow. It is a lover neither of knowledge nor the conversation of mankind. The TSK vision cultivates a knowingness by which the individual can avoid the narrowness of the self and the correspondingly restricted public that it brings along.

The distinction between thinking and identification with the content of thinking emerges more clearly if we investigate the role of the self as the author of a narrative account. Suppose that I tell the story of my life and then am asked to say who I am. If I refer exclusively to the content of the story to identify myself, I am missing the 'I' that is the author (the source of the content). As author the I is transparent, in the sense that nothing identifiable (having content) is the authorial or transparent I. This 'transparent I', the source of the thinking from which content comes, transcends the content of any and every story.

The self is not very open to the arising of new content, for in order to protect itself, the I that insists on its identity must repress, deny, or avoid content incompatible with its identity. This move displaces the transparent I. The transparent I cannot be distinguished in terms of content; it is doing the distinguishing. But to acknowledge this is difficult for the self. As a position or unique bit of consciousness the self is inseparable from specific content. It is likely to find the experience of being indistinguishable at once threatening and incomprehensible.

The presence of bits of consciousness that are distinct from one another—content in other words—is a given, but that content is not the transparent I. One might think of content as the footprints of the transparent I, evidence that the transparent I leaves behind. The transparent I cannot be reduced to this content, and precisely for this reason it always allows for the possibility of new content. As transcendent source it remains transparent, necessarily beyond all content (Kuhlewind 1984, p. 45).

Compare to this distinction Oakeshott's understanding that in the conversation of mankind it is the conversation that matters most. An appreciation of the conversation means that the various voices or modes of imagining, each with its attendant rules and ways, are distinguished and valued. No voice is assigned the identity of the whole, no voice should be displaced, and no voice should become hegemonic. On an individual level, to appreciate the whole conversation entails being free to recognize, to enter, and to exit from the individual voices that together comprise the conversation. This requires that the individual not have an identity that confines her/him to a single voice.

The transparent I serves as the impartial space from which one is free to recognize when the self appears as a fixed position—a closed and unexamined identity. This freedom is also an opening: a space for the arising of content. Just as the various voices contribute to the conversation, so the identities of the self are included in the transparent self (Petranker 1997, p. 268). Just as the whole conversation cannot be reduced to the consciousness of a single voice, so the transparent I cannot be reduced to the consciousness of the individual identified self. Finally, just as the conversation coexists with the voices, the transparent I coexists with the self. In the TSK vision of what is possible, the self is not eradicated, but its relationship to the not-self is opened up. In the experience of these openings the self's usual ways of seeing and operating are enlarged. The cumulative effect of these openings is, as with Oakeshott, the creation of an agent who is more open to and appreciative of the whole conversation, freer in relationship both to its own identity and the identities of others, more compassionate, and more open to a knowledge that is not what the knower knows.

Natural Law and Friendship

Friendship, as John Finnis (1990, pp. 140–41) describes it, is an example of a non-utilitarian relationship—a form of what Aristotle would term pleasure, or what might be called play. "The central feature and good of play," Finnis writes, "is that the activity or performance is valued by the participants for its own sake, and is itself the source of their pleasure or satisfaction." He states boldly, "[T]here is no possibility of understanding the classical tradition of 'natural law' theorizing . . . without first appropriating the analysis of friendship in its full sense."

Note the similarity here to Oakeshott's description of poetry. In poetry the delight is in the play of the images (whether they be in the form of sculpture, music, painting, dance, etc.). In friendship the delight is in the coordination of the well-being of the parties, (and not in what each as a self gets for itself—recall Mansfield's description above of interest as unerotic). In both cases, in poetry and in friendship, the practical self is not in control.

Indeed within friendship it becomes somewhat otiose to refer to individual selves, for in unifying relationships such as friendship the self is eclipsed, at least in part. I would argue that this is due to the emerging of the transparent I. Rather than remain fixated on content (the basis for interests), which precludes freely choosing, the transparent I, with no boundaries or interests to defend, is free to consider the well-being of another or others to be coterminous with its 'own' well-being. Indeed, carried to its full extension the transparent I is one inclusive unifying set of relationships where all bits of consciousness are understood as instances of the whole.

If this way of understanding is right, we could say that as the TSK understanding allows the transparent I to emerge, selves as bits of consciousness no longer identify themselves as bits or positions, but see themselves—and are seen—as knowledge expressing knowledge. The attention of individual agents on knowledge as the arising of bits of consciousness—including themselves as the sources of that awareness—would displace the usual attention on attempting to satisfy substantive desires or interests that have their origin in the self's fixed identification with certain bits of thought content. Individuals understanding themselves, and being understood by others, as spaces for exhibiting knowledge would constitute the space of impartiality, linking traditional concepts of natural law and friendship to TSK and the transparent I.

The seven basic goods of natural law that Finnis discusses—knowledge, practical reasonableness, life, aesthetic experience, play, sociability (friendship), and religion—all entail impartiality, a virtue necessarily present in the position-free space of TSK that makes the transparent I available. The utilitarian positioning of the practical self, on the other hand, stands in the way of experiencing the appeal of the impartial space of the natural law tradition.

In natural law reasoning, the recognition and appreciation of these basic goods does not depend on so-called "impartial" observations about human nature. "Rather, by a simple act of non-inferential understanding one grasps that the objects [the basic goods] of the inclination which one experiences is an instance of a general form of good, for oneself (and others like one)." (Finnis,1990, p. 34) The basic goods are experienced directly and intuitively, rather than through inference.

Recall Oakeshott's description of poetry, quoted above: "Since there is no problem to be solved, no hypothesis to be explored, no restlessness to be overcome, no desire to be satisfied, or approval to be won, there is no 'This, therefore That,' no passage from image to image in which each movement is a step in an elucidation or in the execution of a project" (1991, p. 513). With no project or agenda at work, a clearing appears in which the self-evident and impartial nature of the basic goods appears. In TSK terms, the knower as self entails the filtering screen of the interests of the self, while without the agenda of the self a spacious clearing appears, available for a more continuous unrestricted knowing that is open to the self-evident truth of the seven basic goods Finnis describes. Poetry, natural law and friendship, and TSK and the transparent I meet in the impartial, agenda-free, no-boundary world of knowledge as process.

In particular, it is the practical self that stands in the way of the experience of natural law as an "act of non-inferential understanding." What appears to a self is limited to what is consistent with its interests (to have a self, by the very nature of a self, is to have interests). Locked into instrumental or inferential reasoning, the self is constantly looking for means to satisfy ends that are rooted in its identity. With an identity to protect how can one be fully playful, open to an ultimate order of reality that is not "me" (what Finnis means by religion), and a true friend?

The seven basic goods of natural law appear to the self as "someone's good ideas." But as the transparent I knows, and a TSK perspective on knowledge makes clear, the basic goods are self-evident.

There is a risk here, because pointing to the basic goods through the use of words as symbols, while unavoidable, can encourage a shift from selfless being to ownership of—and identity with—the basic goods as content. The basic goods then become bits of content the knower self can possess and be righteous about, rather than self-evident realizations that appear to someone in the space of TSK and ownerless knowledge. The basic goods of natural law will not present themselves as self-evident 'truths' to the practical self.

Philosophy and the Public

Philosophy is usually understood in terms of content claims: the philosopher is claiming that x and not y is true. Another possibility, however, is that philosophical discourse is a way to loosen the grip of the practical self, so that the philosopher becomes open to the impartiality and playful space of the poet, the transparent I, and the TSK vision. Philip Merlan (1927, p. 423) argues for understanding Plato in this way, suggesting that Plato's Socrates left his interlocutors cross-examined rather than instructed by specific content:

> What [Plato] tried to impart was not doctrines—it was something behind them; and this something was never present to him in such a way that it depended only on his own will to impart it to others. And by retracting every expression, by declaring that the doctrine was not his, Plato was retracting it even for himself. He did so to create something anew, thus keeping his doctrines, keeping himself alive.
>
> Perhaps it is this keeping oneself alive of which Diotima is thinking when she speaks of self-

> begetting. Begetting is the only way in which a mortal being can remain in living and share in the Good forever.[4]

Plato's disassociation from every doctrine suggests that what matters is not content *per se* (doctrine), but the nature of one's relationship to content. For the philosopher as a lover of knowledge it is the arising of knowledge that is loved. The content or doctrine is significant as an episode in that arising. Like the transparent I, the philosopher (at least in Plato's ideal sense) does not identify with thought content, but is a constant clearing for the appearance of ongoing knowledge.

This philosophical stance has implications for the political realm. A political association marks the creation of a public; i.e., a set of unifying relationships by which individuals collaborate and coordinate their activities (Finnis, 1990, pp. 134–60). A public is a form of community, a process of relationship and interaction whereby accomplishments that require cooperation and coordination become possible.

A fundamental issue for every public is membership: who is to be included in the 'we' that the public is? Here identification and positions become relevant. Language, ethnicity, birthplace, family, and so on become the bits of consciousness on which the boundaries of the 'we' are established. We care about those who on the basis of the various bits of consciousness deemed relevant are 'us.'

Friendship is an obvious example of a 'we', a unifying relationship. In a relationship of friendship, friends care about one another. The well-being of A is the well-being of B to some extent, and vice versa. In the *Republic*, 423e–424a, Plato claims that through education (i.e., education guided by philosophers) citizens "will see that all things in the city must be arranged to conform as closely as possible to the proverb that friends have all things in common."

By undermining the blind unthinking hold of identification with bits of consciousness, the TSK vision facilitates more inclusive forms of friendship, rooted in a more playful and impartial consciousness that replaces identity with the transparent I—a way of being that leads to more inclusive publics.

Conclusion

The assumption of the liberal theory of the state is that those who comprise the body politic are practical selves who by and large form no unifying relationships beyond family ties (and possibly ethnic and religious ties); they are intent upon protecting and furthering their practical self-interests.[5] The political association is not united by a common good (see Sandel, 1996, pp. 10–11 and MacIntyre, 1981, p. 112), but by a "free contract among rational individuals for the protection of their natural rights." (Smith, 1999, p. 625)

In contrast, the playful character of the poetic idiom springs from "the self-sufficiency enjoyed by each engagement in the activity and by the absence of any premeditated end." Poetic images are not read as symbols (governed by certain rules and having specific agreed upon uses), but rather (in the words of Oakeshott previously quoted [see p. 133]) "are made, remade, observed, turned about, played with, meditated upon, and delighted in."

The TSK vision gives a way to ground, and in some respects go beyond, this poetic mode of discourse. The transparent I, focused on the arising of content rather than being locked into the substantive content of bits of consciousness, can freely experience the delight of the poetic idiom. Just as the poet is free in relationship to content symbols, because she has no agenda that the content must fit and serve, the TSK visionary is free from content, attending instead to its free arising.

This freedom leads to the formation of more inclusive and unifying relationships. It allows for a public more like an association of friends who genuinely care about each other than the liberal model of a public made up of isolated individuals, coolly negotiating with one another for the protection of their rights.

NOTES

1. The secondary place of the good in political liberalism is a familiar refrain; two examples are Sandel (1996, pp. 10–11) and MacIntyre (1981, p. 112). A related theme is understanding the good as the rule of law, which emphasizes the acceptance of an authority that lays down the conditions individuals are subject to

as they pursue self-interest. See Oakeshott, 1999, pp. 129–78.

2. Oakeshott hints at other voices—the historical and the philosophic—but does not elaborate.

3. Petranker (pp. 262–63) recognizes that this process is social in nature: that in Charles Taylor's terms (1991, p. 33) we are dialogical beings who exist socially in a context of already established content. Generally speaking, it is only within the limitations of our given context or paradigm that we make choices as we interpret reality.

4. Merlan cites the Platonic use of imagery from the world of the blacksmith where the ore aglow and the shape the ore takes cannot coexist at the same time, as the ore would consume itself (1927, p. 430). I suggest thinking of the transparent I as the ore aglow and forms, including the self, as the hardened, cooled-off shape. As the ore aglow, the transparent I cannot be seen with eyes that see only empirical forms; it is seen with the mind's eye, which can appreciate the nothingness or silence from whence form emerges.

5. Oakeshott notes that, "Not only is modern European and American history the story of the growth of the practical voice, at the expense of other modes, it is also a tale of the assimilation of politics to the practical mode" (1991, p. 495). He also notes that the symbolism of the English language in the sixteenth century was "still inchoate" and therefore less resistant to poetry than is twentieth century English (1991, p. 529).

REFERENCES

Balsekar, Ramesh S., 1982. *Pointers from Nisargadatta Maharaj*. Durham, NC: Acorn Press.

Charney, Evan, 1998. "Political Liberalism, Deliberative Democracy, and the Public Sphere," *American Political Science Review* 92, pp. 215–16.

Finnis, John, 1990. *Natural Law and Natural Rights*. Oxford: Clarendon Press.

Hobbes, Thomas, 1958 [1651]. Leviathan. Indianapolis: Bobbs-Merrill.

Kuhlewind, Georg, 1984. *Stages of Consciousness*. West Stockbridge: Lindisfarne Press.

Mansfield, Harvey, 1995 "Self-interest Rightly Understood" *Political Theory* 23, pp. 48–66.

MacIntyre, Alasdair, 1981. *After Virtue*. Notre Dame: University of Notre Dame Press.

Merlan, Philip, 1927. "Form and Content in Plato's Philosophy," *Journal of The History of Ideas* 8, pp. 406–430.

Nozick, Robert, 1989. *The Examined Life*. New York: Simon and Schuster.

Oakeshott, Michael, 1991. *Rationalism in Politics and Other Essays*. Indianapolis: Liberty Press.

Oakeshott, Michael, 1999. *On History*. Indianapolis: Liberty Press.

Petranker, Jack, 1997. "Turning Inward Outward: Toward a Public Self and The Common Good," in J. Petranker, ed., *Light of Knowledge*. Berkeley: Dharma Publishing.

Sandel, Michael, 1996. *Democracy's Discontent*. Cambridge: Harvard University Press.

Smith, Thomas, 1999. "Aristotle on the Conditions for and the Limits of the Common Good," *American Political Science Review* 93 [3].

Strong, Tracy, 1992. *The Self and the Political Order*. New York: New York University Press.

Taylor, Charles, 1991. *The Ethics of Authenticity*. Cambridge: Harvard University Press.

Taylor, Charles, 1992. *Multiculturalism and "The Politics of Recognition"* Princeton: Princeton Univ. Press.

Weintraub, Jeff, 1997. *Public and Private in Thought and Practice*. Chicago: University of Chicago Press.

Causal Knowledge and Causal Emptiness

Ingrid Harris

The TSK 'story' continues with an investigation of the experience of causality. Meditating upon Tarthang Tulku's suggestion that the operation of cause and effect "can be opened to enquiry without placing that inquiry within a causal sequence," I wish to direct your attention to the universal existential structure of causality and its associated causal knowledge. A phenomenological approach of *observing observation* as it operates, as suggested by Tarthang Tulku, will look at patterns of intentional knowledge that escape the confines of conventional intentionality, and will further aim to investigate desire without getting caught up in its flow.

The causal knowledge that permeates our Western-scientific culture is entirely one-sided. This has caused problems. Treating existence as a matter of nothing but objects in external-causal relationships, the determinist-scientistic view has lost touch with our *existential* causal knowledge. It violates our sense of moral reality. It cannot comprehend the significance and the importance of *volition*—the sense of actively and freely choosing a course of action. This paradigmatically human experience of freedom is devalued to the status of mere phenomenon, thereby impoverishing all that is outside the range of positivist enquiry. Let us conceive causality otherwise, however, in a manner that makes it not merely compatible with, but shows its necessity for, human freedom.

We begin with a discussion of the irrevocable "dual citizenship" of human being. Secondly, the scientistic approach is critiqued for failing to do justice to this two-sidedness of human experience. Thirdly, I shall discuss how, from the awareness of the two-sidedness of the lived experience of causal process, there emerges the awareness of the freedom that is our birthright.

I. CAUSALITY AND DUAL CITIZENSHIP

The details of the process from which causal connections emerge and the patterns of intentional knowledge involved in this process lead to an inescapable difficulty; namely, the ambiguity involved in both the language and the experience of *process*. Tarthang Tulku refers to this ambiguity as "dual citizenship," while Maurice Merleau-Ponty made reference to two fundamental "sides" of experience, between which we are constantly in motion. The point is that there are two fundamental ways of thinking and experiencing, which are reflected in two ways of thinking and speaking about, of organizing and reorganizing, the way we inhabit reality. The causal process can also be considered from these two basic perspectives: as a chain of events, so to speak 'observed from the outside', and as a lived process that is *known from the inside.*

The dimension that is accessible from the outside is the *objective body,* which is spoken of as a system of bones, muscles, and flesh. Then there is the inside, which Merleau-Ponty calls the *lived body,* the living experience *by which there are objects.* The phenomenological inside is not simply an object not ordinarily accessible to others, such as the *interior* cavities of the body that contain the *internal* organs. The lived body is "the feeling that one feels, the seeing one sees[; it] is not a thought of seeing or of feeling, but vision, feeling, mute experience of a mute meaning." (Merleau-Ponty, 1968, p. 249)

There is no abyss between these two sides of experience. As the dual-citizenship metaphor aptly expresses, we constantly move from one to the other, from the external to the internal and *vice-versa.* Directed awareness, i.e., *reflection,* upon the two sides of experience and our freedom of movement from one to the other reveals the ability to make this shift intentionally. And attention directed to the moment of the shift makes evident the fundamental indeterminacy of lived experience. At the moment of transition, 'we' are neither object nor subject but somehow both and somehow neither (compare LOK, p. 369 [exercise 45]). This moment overflows the boundaries of the conceptual categories conventionally used to organize experience. By disrupting these categories, it lets us see their relativity and their contingency. The moment is like trying to touch the finger that is touching another finger, or trying to see ourselves looking as we look in a mirror. A reflective and reflexive process occurs, in which we distinguish ourselves from objects and, nevertheless, see ourselves as objects (Merleau-Ponty, 1962, p. 93).

Reflecting upon the formation of the 'internal' and 'external' sides of human being, and the movement from one to the other, relaxes the tendency to act as a subject grasping objects. Pausing in the space-time between the movements of subject and object opens access to *causal knowledge*. We see objects emerge from background by means of the positing of identity, ready to operate in temporal sequences of causes and effects. We see how causal knowledge is an *active engagement* of human being with the world. Pursuing the causal question, "How have 'I' seen 'this' arise?" (LOK, pp. 359–60), we gain access to the *intentional* process of causal thinking, in which conventional subject-object patterns and habitual conceptual structures are *set up* and *put together*.[1]

Causal knowledge is at the heart of human being. As David Hume observed in his *Treatise on Human Nature*, to try to anticipate outcomes and to improve prospective outcomes is central to the human condition. Hume noticed that from the constant conjunction of objects or events, there emerges in first-hand or lived experience an *intentional* association among events, one that is characterized by a 'sense' of necessity. As I shall try to show, awareness of the intentional nature of this process liberates us to choose for ourselves the form of identity and hence the form of reality within which we will operate.

Causality and Contact

Causal knowledge bears witness to a *living contact* between self and world. As witness, it gives evidence of the self's dual citizenship. This is illustrated whenever we move freely across the boundary between subject and object, between self and other. As Tarthang Tulku (LOK, p. 151) observes:

> It is one of the puzzling facts about the self that it also has object-like existence. Through 'its' body, the self interacts with other objects, in ways that seem better described by the 'space', 'time', and 'knowledge' of a 'space-centered' objective reality.

The reason this is puzzling is that we do not experience the body only as a 'thing' among other 'things'. We find in lived experience an 'original' of the body that is not 'out there' among things but is always on 'this side' of all things seen.

Dual citizenship means that these two sides of experience are 'always already' in contact. We can, for example, examine the experience of touch from inside and outside; we can actively *feel* another part of our bodies or we can passively *be felt* by another part of our bodies. Not only are the two sides in contact, but there can be direct communication between them. Every time we speak, our voice is both 'heard' and 'uttered', and each of our fingers is a fluidly reciprocal 'space' of felt-feeling.

The same duality holds for seeing or feeling the other. First of all, to see or feel my body as an object is essentially the same as to see or feel the other. Secondly, whether we observe another person, a tree, a rock, or our own limbs as objects in space among other objects, it is always by way of subjective experience, which itself cannot be seen or touched. To conceptualize this immediate and direct contact between subject and object, between self and world, we could think of the sides as reversible—the 'inside' (the mind, the domain of the self) as the 'lining' of the "outside" (the sensible, the 'objective' world) and *vice-versa*. Along with Hume and Merleau-Ponty, we may conceive of the portions of experience that count as objective as all having a subjective lining, which is not detachable.

Metaphorically, the mysterious 'stuff of experience acts like a möbius loop, a simple loop of paper with a single twist that shatters our everyday conception of space (see next page). Although two 'sides' are visible and touchable at any point along the loop, 'inside' and 'outside' are not ultimately distinguishable. Although the twist is indisputably 'there', right in front of us, nonetheless whenever we 'grab' it between our fingers, it somehow 'escapes' our grasp. It remains enticingly and indisputably *elsewhere*. The two 'sides' of human

being are related in this manner. When 'this side' of experience reaches out to touch the 'other side', it never quite coincides with the 'out there'. Instead, its grasp is always completed in a 'there is', in an 'other' that eludes any attempt to take hold of it (see Merleau-Ponty, 1968, p. 261). When we try to continue along one side, that side inevitably, but often imperceptibly, shifts to the other. We could conceive knowledge as emerging when the two sides of experience interrogate each other in a reciprocal process of creating reality.

The möbius strip metaphor permits an intuitive understanding of the synchronic *move* from side to side of experience in a way that preserves both sameness and difference. As well, it is adequate to the diachronic dimension of experience. As much as we need to think objects as wholes that have an objective unity, we need also to recognize how they are different from one time to the next.

Causality exemplifies this interweaving of the sides of our experience along an open 'border'. Causality in scientific terms is *partes extra partes*, that is, 'objects' interracting with other 'objects' independently of any 'subjective' dimension. What gets lost is the first-hand experience of causal relationships as they emerge. As

Hume notes, from a constant conjunction of objects or events there emerges an intentional association among those events, one characterized by a sense of necessity. The subjective event (the sense of necessity) is characteristic of the objective event (constant conjunction).

Say the fluorescent light over your sink does not work when you turn on the light switch. You happen to notice that when you slam a cupboard door, the light comes on. The next time you want to turn on that light, you try slamming the cupboard door; the light comes on. After repeated experiences of the conjunction of slamming and lighting, the idea of lighting is inferred[2] from the action of slamming. The subjective event, i.e., inference, emerges out of (along with) the constant conjunction, i.e., the objective event.

As A. Michotte, who experimented on the perception of causality, concluded (1963, p. 263), the fact that the sense of necessity inherent to the inference is a *feeling* does not mean that it can be detached from the causal event as 'purely subjective'. *The phenomenon (feeling) is characteristic of the event.* Just as in that moment long ago when someone discovered that striking two stones together regularly produced a spark, the entirely unlikely *association* of the phenomena of slamming the door and the light going on emerges out of the living experience of constant conjunction. Many of the cause-and-effect associations that were important to cultures of the past have been lost to us or are believed to have been superseded by current styles of association. The *structure* of causal knowledge, however, has remained constant through time and space.

We cannot but rely upon that constant in order even to have something that counts as 'experience'. In this sense, 'causality' is a sort of *meaning* that emerges among the innumerable patterns of experience. It serves the crucial purpose of distinguishing what is *reliably connected,* what we term reality, from what is not. Causal relationships found the practical knowledge we depend on in our everyday lives.

II. CHALLENGING THE OBJECTIVIST MODEL OF CAUSALITY

Hume and Merleau-Ponty both argue that the objective mode of thought, which is central to natural science, is not prior to perceptual experience but is, rather, the outcome of and the sequel to perceptual experience. Objectivism is an abstraction from lived experience: a cultural artifact (see Gross, 1990, p. 7). If we forget this, we risk losing contact with first-hand perceptual experience entirely.

This is just what happens when we think scientifically about causal relationships. We tend to foreground the constant conjunction of objects (objective) at the expense of the inference between them (subjective). As Michael Polanyi (1962, p. 59) noticed: "Externality is clearly defined only if we can examine an external object deliberately, localizing it clearly in space outside."[3] In doing this, we leave the fact that objects are always the objects of *someone's first-hand experience* out of the picture. For example, we can explain the visual perception of a block of marble as produced by innumberable stimuli impinging upon our retinas, but this account presupposes the actual living process of coming to perceive the block of marble *as* (meaning) "a block of marble" in the first place.

Proceeding scientistically, we interpret our own perception as a product, an 'effect' produced by external objects externally related to our body as object. What gets forgotten is that abstract notions, which allow us to tell a theoretical story of our experience, presuppose lived perceptual experience as the context in which and out of which they receive their meaning. The objectivist approach fails to acknowledge that the notion of, e.g., a visual *sensation* is an artifact of a certain *method* or style of thinking about lived experience.

A scientific notion such as *sensation* belongs to a picture of the human organism as a particular kind of 'object' (for example, a physical system undergoing stimuli). To make this picture work, experience itself is packaged for the purpose of quantification: 'quanta' of energy, 'moments' of time, 'bits' of data. Once distinctly separated, such 'bits' can be substituted for one another and are thereby made amenable to *ownership* and *control*. Initially posited to help make sense of experience in a certain way, such hypothetical constructs take on a life of their own.

Such post-mortem autopsies, or maps, tend to get mistaken for living reality. Like Narcissus, we fall into our own images. A paradise of living awareness of immediate relationships with the world gets overwritten by the abstract concepts that were intended to aid us to understand certain dimensions of that very awareness or experience. Materialism devalues first-hand experience to the status of mere phenomena, unreal in comparison to the 'real' reality of its own theoretical frameworks.

Like every other system of knowledge, scientific materialism asks certain questions of the world, questions circumscribed by certain interests and certain presuppositions. Consequently, as Hans-Georg Gadamer (1990, pp. 369-79) points out, the world gives answers only to those questions. The questions that are posed in scientific methodology characteristically subordinate first-hand ways of knowing to the replication and dissemination of a prescribed belief-system. But that is not the only way to conceive and to approach the world. Consider an example.[4] Suppose we say that the sun rises in the morning. That knowledge is incorrect from a scientific perspective, but it is correct in everyday terms. The statement can refer to either of two different frameworks, systems based upon two different frames of reference, both of which we rely on at different times to interpret experience. As long as the frameworks are not in immediate conflict, we can maintain them both comfortably.

In fact, in everyday life we easily maintain not only these but many different conflicting systems as a matter of course. Rarely do we notice how often and how easily we shift from one theoretical frame of reference to another—as easily as we shift from one software program or one television show to another. In our *practice,* we are remarkably flexible and pragmatic, yet in our thinking, we often fall into the trap of thinking that the world must *be* as a particular theoretical framework circumscribes it.

That, unfortunately, is what has happened with causality. For instance, if I decide to answer an email, a causal determinist might 'explain' how my decision came about by pointing to physical events in the brain that are caused by other events and so on. That explanation individuates my 'self' as an 'object' constituted by other 'objects' in *partes extra partes* relationships, all of which unfold in accord with inexorable laws. Under that view, my writing the email is an unavoidable response to prior conditions. There is no possibility of describing it as a freely chosen act.

Applied to causality, the objectivistic model individuates things in such a way that every thing and every event becomes a product of objective causal relationships. Whatever human beings do is thought to be entirely explainable in terms of prior causal relationships. For instance, science understands mental alertness as an object, i.e., a brain state. It may then link that state with another objective fact; say, absorbing caffeine. Experimenting, the scientist may find that drinking coffee or tea and mental alertness (as experienced first-hand by the individual) are constantly correlated, and further, that the reported state of mental alertness is constantly correlated with a 'state' of brain activity (as measured by some electronic device).

The next move is to claim that the brain state explains the mental alertness (i.e., the brain-state is real because objective and is therefore responsible for the mental alertness, which is merely phenomenal because subjective). But this move is justified only if we buy the idea that what is being registered on the instrument that measures brain activity is more real than what we are presently experiencing. Otherwise, we might just as plausibly claim that the (living experience of) mental alertness explains the changes in the electronic device that are collectively termed "brain state;" that is, that the lived mental alertness is what causes the instrument to register what it does. (This alternative would be most readily available in the case where the scientist is measuring his or her own brain states and thus has immediate access to both kinds of data.) A more neutral description would be that "mental alertness" and "brain state" are two different ways of understanding the same event, one pointing to first-hand experience and the other to the read-out from a scientific device.

Causality conceived as a mechanical force cuts off the sense of living contact with experience. The universe becomes an unbending totality. Every choice, no matter how trivial or insignificant, is preprogrammed by the 'system', and we fall into 'deterministic hell'. Trying to escape is likewise an act determined by prior conditions, which confirms that escape is impossible.

When we *accept* a system as a totalistic view, we become bound; we are born(e) into a world. On the other hand, rejecting such views means that any system we devise loses the power to bind us. We are free to ask different questions, and thus to get different answers, ones that reveal different dimensions of the world. When we look at causality as lived experience, the alternative that science excludes, we at once discover dimensions not amenable to the quantifying methods and externalizating of the objective scientific approach, dimensions for which other styles of thinking are more adequate.

Say that in the earlier example of writing an email, a friend who drops by suggests that I am writing an email because I am putting off doing something more pressing (editing a difficult manuscript). This kind of causality addresses the question of motivation, which the reductive, scientific sort of explanation leaves untouched. Unlike the scientific view, which individuates physical events, making them the *subjects* of discourse, in this view, the subject of discourse is the whole person: the 'one' who is writing the email, who has motives and acts in terms of them, and who wants to understand them better.

Choosing a level of individuation means making a choice about meaning. Normally we make the 'usual' choice, and do so unthinkingly. But there are on occasion encounters with what is radically other, experiences of deep existential import, when the chaff of intellectual systematizing gets blown away and we are left face to face with the fact that we do choose how to individuate and define fundamental meanings (such as 'mind' and 'body', 'self' and 'other') so as to render life as a whole meaningful. On such occasions, we cannot escape the knowledge that we choose, whether that choice is active (e.g., a deliberate opting for one way of defining things *as* of a certain sort), or passive, a simple acceptance of the definitions offered by culture and convention and the style of thinking/acting that accompanies those definitions.

Whichever is the case, the fundamental individuations we make sound an overture to the attribution of causal relationships. It is the intentional *act* of individuation of some sort of whole (e.g., 'self' or 'other') from a background within which it is situated that initiates the action of taking a position. 'We' and other 'individuals' emerge from fundamental acts of individuation as certain sorts of individuals, and we come to be causally connected with the world and with others in ways that are entailed by that initial individuation. Each initial act of severance, each choice of individuation, initiates a different 'universe'.

Difference does not entail hierarchy. Any system of thought or interpretation requires initial individuations, but all these analyses occur on the same level.[5] Claims of relationships of 'superior' or 'inferior', based upon being more or less close to 'ultimate reality', are not credible. Causal explanations, whether they refer to scientific objects and forces, motivations, or any other kinds of individuation, are simply stories we tell ourselves about the world and about ourselves. They all have the same structure and in that sense are equivalent. This is the significance of Hume's observation that all our knowledge of causality comes down to knowledge of constant conjunctions. Since how we individuate events determines what sort of patterns of conjunctions are relevant, the precise nature of the story depends upon our intentions. When we see this, all stories become, so to speak, 'transparent'. Causality itself becomes 'empty'.

Recognizing that causal relationships are stories; i.e., discovering the emptiness of causality, does not mean that we can simply dispense with causal language or label causal relationships as merely relative. After all, causal connections in the everyday world are responsible for suffering. The failure to think causally about the likely outcome of one's actions more often than not has suffering as its consequence. To go about reducing all forms of knowledge to nothing but stories or interpretive constructs would leave us with the same experience we have now, except that we would have traded unquestioned realism for unquestioned nihilism. In reducing the value of causal knowledge to illusion, to the meaningless void of a conventionally understood 'empty', we would have reduced our own value in the process.

III. REGAINING CAUSAL EXPERIENCE

Understanding the causal process in a more open way permits us to see how causal relationships emerge from distinctions that are made by human beings for human purposes. Because these distinctions are flexible and fluid, the causal process is fundamentally open. Causal relationships inform us of reliable associations among phenomena. They permit us to distinguish the regularities that transcend individual peculiarities and move towards universality. Through identifying such relationships, we are able to distinguish what we imagine or experience subjectively from what is real, certainly a distinction of the utmost importance.

The role of faith in the formation of causal relationships is revealed by Hume's deconstruction of causality. Hume (1978, p. 166) (Book I, Part III) observes that from the constant conjunction of objects or events, there emerges an association among them, characterized by a 'sense' of necessity. Before we can identify causal relations among objects, however, we must first individuate isolated objects, which can then be categorized in terms of similarities and differences. But these individuated structures, like causal relationships themselves, have no independent foundation. They grow out of the tendency toward objectivism, coupled with habitual ways of thinking. They are not 'wrong', but they do tend to occlude awareness of the interdependent origination of objects. For instance, when I see red, the idea that 'I' exist *isolated* from 'redness-in-itself', which I then somehow 'see', is an abstraction based upon making initial dualist individuations between 'self' and 'world' and categorizing experiences on that basis. Following the logic and grammar of the sentence, "I see a red thing," this method treats causality as a matter of relating isolated atomic components. But we have already come across an alternative in the metaphor of the möbius loop. This model allows us to say that the object (the seen) is immersed in the subject (the seer), and the subject is immersed in the object. From this perspective, there is no gulf, just "I-see-a-red-thing."

If we look directly at the *operations* through which objects are identified and causal relationships assigned, rather than the assigned identities, the starting point is the unique experience. In the 'perception' of touch, for example, novel living structures of knowledge emerge through the interactions of space and time. When 'I' touch a 'stone', a 'communion' occurs that did not exist before and will not exist after this present moment, when this unique space 'substance' (e.g., *this* stone) presents its tactile being to my senses. In this living communication between 'opposites', a transformation of both takes place.

This communication, an immediate unique experience, provides the basis for moving toward the universal. Once we have had the lived experience of touching the stone, our bodies know *how* to anticipate the touch as well as know *that* this stone has its particular texture, hardness, weight, etc. Our eyes and our hands come to *know* that changes of place will produce sensible responses according to expectations. By embodying knowledge in this way, we come to feel in anticipation the countless mass of more detailed perceptions yet available to our senses. Only when we have received an infusion from this other space and this other time that is the stone can we then anticipate the experience of other stones in the future.

Seen in terms of operation, this interaction is the foundation of all causal knowledge. Space and time are not only good friends but generous benefactors. Freely and continuously they offer their abundant gifts of causal knowledge, permitting us to learn from whatever happens. Causal relationships, then, may be spoken of from both 'sides' of human being. They are the way that we make contact with what transcends our subjective immanence. And they are also the way in which what transcends our subjectivity touches us in the present, immediate encounter, granting us gifts of lasting value.

Knowledge and Anticipation

We regularly develop theories about what causes what and what will come next. Any such theory arises out of and is abstracted from the complexity of experience. William James (1977, p. 231) was pointing toward this distinction when he wrote:

> Experiences come on an enormous scale, and if we take them all together, they come in a chaos of incommensurable relations that we can not straighten out. We have to abstract different groups of them, and handle these separately if we are to talk of them at all. But how the experiences ever *get themselves made*, or *why* their characters and relations are just such as appear, we can not begin to understand.

The real causal knowledge is located not in the theory but in the experience. Let us take a concrete example. In the age of colonization, indigenous populations that had never dealt with guns before the arrival of Europeans sometimes believed that amulets would defend them against the invaders' bullets. This belief proved false. But it was not their causal knowledge, i.e., the accumlation of past experience of the constant conjunction of amulets and protection, that was at fault.

Rather, what failed them was their *anticipation* that amulets would protect against the alien weaponry, just as they had protected (let us suppose) against spears. This was a theory. The indigenous peoples that held it had no experience upon which to base their anticipation since they had not yet tried amulets in this way.

Of course, an outside observer might say that this is nonsense, that the amulet did not work against guns *because* it did not in fact work against spears. There was no real causal connection; i.e., the conjunction of amulet and safety in battle is not universal. We may wish to say it in this way. But there are other alternatives. It may be that amulets were an aid to concentration or to summoning courage or martial skills and in that way assisted in hand-to-hand or at least close combat. Consider how prayer or meditation has been recognized to be helpful in dealing with, alleviating, and even sometimes curing life-threatening diseases.

It would take us too far afield to go into much detail here, but it must be recognized that an amulet is not simply the visible 'object', which somehow in and of itself does the job of protection. Surrounding the 'amulet' is a complex cultural web of beliefs and practices (i.e., technology, theory). The amulet is, so to speak, merely the visible 'point' of the web of meaning in which the amulet-wearer lives. To say that amulets never 'really' did anything because bullets can kill people wearing them is like saying that modern medicine does not "really" do anything because it does not protect against viruses or because it is defeated by some serious illnesses for which prayer and meditation have arguably been remedies. Hume's point is that we never really get access to the so-called 'real' connection. We are limited to observing the constant conjunctions and our anticipation is formed on that basis.

Causal explanation, then, arises within an entire causal nexus, a 'way the world is'. Recall the unlikely association between slamming the cupboard door and the light coming on. It makes sense to us because we can fill in the gaps with our knowledge of our particular technology; i.e., the theory of electricity, wiring, etc. In the absence of such a context, we would most likely see this as a 'magical' connection.

Nonetheless, causal relationships do not arise solely from observing the association of events. Causal relationships also presuppose that the future will resemble the past. Because they rely on this presupposition, which cannot be proved by any arguments, causal relationships are dependent upon a form of *faith*. The point to bear in mind is how causal thinking operates. Because abstractions generalize the innumerable particularities of experience, anticipations always 'run ahead' of actual experience. Anticipations rely upon the presupposition that the future will resemble the past, and this means they can give rise to false predictions. Consider another example: A very young child warms herself at a fireplace. The pleasurable experience of warmth makes her associate the fire with that pleasure. She anticipates that moving closer will increase the pleasure. When she experiences pain when she touches the flames, she finds her anticipation to have been erroneous.

This example introduces a new point. The child's idea of fire, and of its causal relations, is modified by her *experiences of the objects themselves.* As Hume repeatedly emphasized, causal relationships are established only through lived experience and cannot be foreseen—no matter how careful the scrutiny—by trying to find some unchanging and unchangeable 'energy' or 'power' in the so-called 'cause' from which the effects might be *deduced* as unquestionably contained, prior to the effect's having come about. Concerning this point, Hume (1989, p. 29) wrote:

> A stone or piece of metal raised into the air, and left without any support, immediately falls: but to consider the matter *a priori*, is there anything we discover in this situation which can beget the idea of a downward, rather than an upward, or any other motion, in the stone or metal?

Ordinarily the distinction between causal knowledge and anticipation has little practical significance. We *anticipate* that the future will be like the past and this act of faith forms a reasonably reliable basis for the practical efficiencies of everyday life. As a matter of course, we draw upon past causal knowledge (the outcome of a long course of informal experiments) to plan for the future, and this seems wholly appropriate and beneficial. Difficulty arises only when we make the metaphysical leap to the conclusion that something 'inevitable' underlies all these connections, something in principle incapable of being experienced, yet attainable by objectivist theories, that weld everything in the universe into a rigid, unbending unit.

Not all constant conjunctions are in practice inevitable and not all causal knowledge allows us to anticipate accurately what will happen. The lives of yogis and saints, such as Milarepa, who challenged even the most accepted causal relationships, show that widely agreed upon regularities are not entirely incontrovertible. But even those of us who accept that fire always burns and water drowns cannot say that what is habitually so will always account for what happens next. The 'same' extreme conditions, for example, give rise to radically disparate behavior: decent and indecent. Survivors of concentration camps or prisoner of war camps in any time or place remember with gratitude and reverence the men who walked through the huts comforting others, giving away their last piece of bread. Such individuals are the exception, but their enduring legacy is the proof that we have the freedom to choose our attitude in any 'given' set of circumstances. The belief that every action is inevitably determined by prior conditions is a hasty generalization from a small set of constant conjunctions over which we believe we have no authority.

IV. CAUSALITY AND FREEDOM

We have to a greater or a lesser extent become blinded by the objectivist view of causality, but as Hume points out, the true paradigm of causal relationships is to be found in *volition*. In voluntary action, we *feel* the connection between our attempts to act and the results of our actions. Objectivist views of causality have little to say about this. But if we focus on causal *experience*, we see that this felt sense of volition is in fact the fundamental *lived inference* from which the very meaning of 'causality' stems. Any model of causality that reduces volition to the status of 'merely' phenomenal or subjective has lost touch with precisely the sense of connectedness and of power to bring about a result that is characteristic of causal experience. And this means that it has also lost touch with the living potential for transformation within the present situation, for creating a self as an original starting point of events rather than merely a 'transmitter' of a push from elsewhere.

Restoring volition to the center of causal knowledge answers the objectivist claims of determinism. Determinism overlooks the fact that *we forge* the connections in our lives by our actions—including the actions of mind that choose what counts as a cause and hence what is conceived as a connection that enslaves us. Getting clear on this truth leads us to the view put forward by Tarthang Tulku (DTS, p. 68): "[C]ausality itself becomes more free: not dictating in accord with an automatic logic; not arranged by fiat; not compelled in the nature of things."

The emptiness of causality is not the emptiness of no cause/no effect, spaceless and timeless. It is the emptiness of being non-binding. We are responsible for conditions because *we have a choice* in how to meet them. By meeting them in one way rather than in another, we choose the consequences (effects) of the initial conditions (causes). The knowledge of how objects emerge from their background by means of the fundamental positing of identities, of how they operate in temporal sequences of causes and effects, makes all the difference in how we individuate ourselves. Do we choose to identify ourselves, for example, as "matter in motion?" As "living energy?" As "evolving Bodhisattvas" or "embodied space, time, and knowledge?" The nature of the world, the self, and their interrelationships are born based on how we individuate subjective and objective.

Far from being fixed and inevitable laws invisibly and inexorably ruling everything, causal relationships are a manner of communication between the world and ourselves. Through causal relationships, the world continuously constitutes and supports the flexible and fluidly structured fabric of reality. We can come closer to this understanding by questioning the habitual structures of meaning that give rise to the idea of causal 'necessity'; by doing so, we can learn to imagine the unimaginable. Instead of letting ingrained habits of thinking overwrite our living experience, we can look closely at the ongoing process of causal knowing. We can discover that the intricately integrated and infinitely variegated web of space and time *allows but in no way determines*. It shapes a structure of knowledge *within which* the mind works.

Identity and Knowledge

Each person in each moment is a unique and open-dimensioned *style* of integrating and transforming time, space, and knowledge. Each individual continually participates directly in changing the dynamic structures and patterns of existence. We are free because of and in the midst of conditions, not in spite of them. We discover the wealth of causal knowledge, incorporated into each conventional identity.

The structures of identity made visible by causal knowing give us, so to speak, a grip upon the world. The more causal relationships we understand, the freer we become to act so as to bring about our intentions. This can happen because causal knowledge takes us directly to the level of volition, which permits us to change our attitudes and thereby to become capable of transforming our being.

Activating this kind of freedom depends upon recognizing that our senses and minds take an active role in shaping the world into which the self emerges. It is the intentional *act* of individuation of some sort of whole (e.g., self or other) from a background within which it is situated that initiates the action of taking a position.

Causal knowledge lets us watch a sequence of 'I's emerge in experience, noticing the associated 'feelings' and 'desirings' of the 'read-out' of any particular moment (LOK, p. 367). The boundary between self and world is revealed as a *contract* as well as a *contact* with knowledge, as an open space that is fluidly structured and amenable to change.

Consider, for example, one of the circles of suffering, the 'natural' tendency toward pride of place. When intensely associated with the idea of self (i.e., of ownership of one sort or another), the pleasure that is brought about by the beauty of surroundings is not appreciated as it is; instead, it quickly transforms into a sort of vanity. When ownership is threatened, the enjoyment or vanity is even more quickly transformed with the aid of grasping into strongly negative emotions such as anger and frustration.

The same sorts of associations occur concerning property, riches, climate, antiquity of name, honors, etc. Even when ownership is set at a distance; say, when the beauty 'belongs' to a relative or friend, the 'gravitational' force (LOK, p. 274) that transforms enjoyment almost immediately into vanity remains strong. The resulting emotions themselves become causes for the actions that follow, and these actions in turn establish the basis for future recurrences. Tarthang Tulku observes (LOK, p. 374): *"Whatever is thought is so; whatever is acted on establishes the foundation of action."* Through such patterns of possession, the world loses its living luminous openness and becomes ever more confining, until 'identities', 'things', and 'events' finally *weld* together as the prison-bars of the determinist hells.

Causal knowledge redirects the energies that shape the determinist hells. It lets us attend to the way in which the objects of desire and aversion 'cause' us to act. Consider the example given above: My friend observes that whenever I have something pressing to do, I break off to compose an email message. Her observation is relevant to me as a 'person', because I have some intention or concern about my motivations. Hence I may reflect on whether this association holds.

I observe my own internal events as well as the events accessible to everyone else, so as to decide whether my friend's hypothesis is warranted. Is there constant conjunction? Is there at least sufficient regularity to count as a causal sequence?

Focusing on causal structures in this manner reveals their openness, their susceptibility to change. This permits what Tarthang Tulku (LOK, p. 349) calls "a 'being-in' that is not confined by the specific manifestation of the 'field'." When we understand that there is no inevitability to even the most familiar and oft-repeated regularities, we become free to work with causal sequences in new ways—to practice what Tarthang Tulku elsewhere (DTS, p. 70) calls "zeroless responsiveness."

One powerful methodology for activating this understanding is to reverse the tonalities on particular feelings and desirings (LOK, p. 341). By intervening in such tonalities and the causal thrust they generate, changing the direction of their gravitational pull towards action, we can negate whatever is established while it is in the act of being established. By *using* causal knowledge in this way, we reveal both its availability and its power to transform our causal reality.

Invested with this authority, the self is freed from the tyranny of 'the past'. The two-sidedness of experience, the ability to move from inside to outside, means that we can see ourselves as both the products and the producers of history. The flexibility of individuation means that in the pursuit of understanding we can generate innumerable accounts in which we play both roles. This does not mean that either 'we' or 'the past' dissolve into a maelstrom of randomly flowing energies. Instead, as Hume's work makes evident, 'human nature', like 'the past', is an idea, discerned as regularities pervading the experiments of life. The 'past event as embodied in the present because of the choices that were made at the time' is not anywhere near all there is to the matter.

Here is an experiment: Try for a moment considering whether someone has or has not harmed you by some action. Run through the sequence of events under both interpretations and see how your relationships with that person would be transformed by actions stemming from the one view or the other. The point here is that when one settles upon an interpretation, one's actions flow from it. For example, given the same instigating actions, one might equally well accept an apology or begin litigation. One's actions, as it were, give the drama of the whole enterprise its last act, which in turn shapes the drama as a whole. For once a course of action is initiated, it reinforces the interpretation that led to it. Then the world settles into that sort of world.

This process, which produces results that seem so fixed, could become open again. Any course of action is itself a source for causal reflection. The sort of world we have is dependent upon the sort of individuations made (by habit or by deliberate choice) by each of us at each and every moment.

It follows that we are always free to change the curriculum of our lives. Causal knowledge gives us the alchemy that lets us transform causal necessities. When we take up the reality of causal knowledge and causal emptiness, we assume responsibility for our relationship with time, space, and knowledge. We achieve the necessary goodwill to enter into full partnership with the world in the great work of creating a new way of being.

NOTES

1. The phenomenological concept of *intentionality* marks the feature of '*aboutness*' or *content* that is central to phenomena. In other words, it marks the fact that there are not merely phenomena 'given' in some absolute sense. It points out the fact that *meaning* is central to the very fact of phenomena and cannot be dismissed as something 'subjective' or 'psychological' that is 'added on' to the 'bare objectivity' of objects.

2. It is interesting that the English word 'infer' is derived from the Latin *infero*, to make or produce. The 'connection' between slamming and lighting is 'produced', so to speak, by the constant conjunction of those events. The subjective experience (event) of connection is produced by the objective events.

3. Gross describes scientific 'discovery' as rhetorical invention: "Why redescribe discovery as invention? To discover is to find out what is already there. But discovery is not a description of what scientists do; it is a hidden metaphor that begs the question of the certainty of scientific knowledge . . . . To call scientific theories

inventions, therefore, is to challenge the intellectual privilege and authority of science . . . . If scientific theories are discoveries, their unfailing obsolescence is difficult to explain; if these theories are rhetorical inventions, no explanation of their radical vulnerability is necessary".

4. My thanks to Jack Petranker for his helpful and insightful contributions, which I have incorporated to help to clarify this and numerous other passages.

5. See Hume, 1978 (Book I, Part III, Ch. 14): "All causes are of the same kind, and in particular there is no foundation for that distinction, which we sometimes make betwixt efficient causes, and causes *sine qua non*; or betwixt efficient causes, and formal, and material, and exemplary, and final causes. For as our idea of efficiency is deriv'd from the constant conjunction of two objects, wherever this is observ'd, the cause is efficient; and where it is not, there can never be a cause of any kind . . . . The same course of reasoning will make us conclude, that there is but one kind of necessity, as there is but one kind of cause, and that the common distinction between moral and physical necessity is without any foundation in nature."

REFERENCES

Gadamer, H-G., 1990. *Truth and Method.* New York: Crossroad Publishing.

Gross, A., 1990. *The Rhetoric of Science.* Cambridge: Harvard University Press.

Hume, D. 1978 [1739]. *A Treatise of Human Nature*, second edition, ed. L.A. Selby-Bigge. Oxford: Oxford University Press.

Hume, D. 1989 [1748]. *Enquiries Concerning Human Understanding and Concerning the Principles of Morals*, second edition, ed. L.A. Selby-Bigge. Oxford: Oxford University Press.

James, W., 1977. *The Writings of William James*, ed. John McDermott. Chicago: University of Chicago Press.

Merleau-Ponty, M., 1962. *Phenomenology of Perception* (London: Routledge & Kegan Paul).

Merleau-Ponty, M., 1968. *The Visible and the Invisible: Followed by Working Notes*, ed. Claude Lefort. Evanston: Northwestern University Press.

Michotte, A., 1963. *The Perception of Causality*. London: Methuen.

Polanyi, M., 1962. *Personal Knowledge: Towards a Post-Critical Philosophy* Chicago: Univ. of Chicago Press.

A New Framework for Physics

Maxim Osinovsky

Our science, both blessing and curse of mankind, was laid down during the sixteenth and seventeenth centuries in the course of a fierce struggle between emerging scientific knowledge and medieval patterns of thought. This older system of thought included a natural philosophy based on a view of the world as a harmonious whole, or organism, governed by a Deity and managed through the mediation of a realm of Platonic ideas. Thus it was believed possible to decipher natural phenomena through a correct interpretation of the Holy Scripture, allegedly containing a full exposition of the Divine plan for the world; or else through the careful manipulation of abstract categories such as essence, substance, attributes, purpose, and intention.

Having gained victory over medieval superstitions, science cast away natural philosophy. What is more, it refused to use the way of thinking based on descent from some embracing vision to particulars. This deductive method gave way to the inductive method of gathering facts, formulating theories, and applying theories to emerging problems.

The new program for the development of science was first formulated by Sir Francis Bacon, who considered Nature as something to be conquered and controlled by using the power of inductive knowledge. This program was adopted by the most prominent scientists of that time, as in Newton's arrogant statement in favor of pure facts: *Hypotheses non fingo* ("I do not invent hypotheses.")

In denying vision-based methods and concepts, science also denied considerations of purpose, the good, the right, and the beautiful. Lacking a scientific basis, religious and ethical ideals were undermined. Over more than three centuries science and technology triumphantly progressed from success to even greater success: steam engines, electricity, radio, television, lasers, A-bombs, rocket ships, genetic engineering.

By the second half of the twentieth century, it became clear that this development was fraught with unpleasant consequences. From the social standpoint, the unimpeded progress of science became a threat to mankind's very survival. From the personal standpoint, scientific knowledge brought about frustration and loss of orientation. It explained "what" and "how" very well, but did not speak of "why" and "for what purpose," claiming these questions not to be scientifically relevant. But the continuing accumulation of knowledge without any purpose leaves some vital aspirations of human beings unfulfilled: the deep craving to know who they are, to know their origin and fulfillment, and the nature of reality. It seemed that science would never answer these questions, being unwilling even to consider them.

At the same time, science found itself by the middle of the century confronting with ever greater urgency the necessity to bring the logic and methodology of science into accord with new discoveries involving layers of reality lying so deep that they cannot be contacted by the ordinary senses. These attempts are now underway, and there is not yet a coherent, all-embracing, and generally accepted interpretive scheme. Recent theories invoke parapsychology, the hypothesis of formative causation (Sheldrake), the implicate order (Bohm), the holographic paradigm, and others, attempting to restore discarded and long-forgotten concepts such as purpose, wholeness, aliveness, the hidden order of the universe, etc.

In the context of such efforts of Western men and women to know themselves together with knowing the external world, we can better understand the significance of the TSK vision, which appeals to our innermost urge to know these three most fundamental facets of our experience and reality (time, space, and knowledge). In contrast to philosophical or scientific treatises devoted to the same subject, Tarthang Tulku invites the reader to experience for himself or herself three levels of space, time and knowledge, roughly corresponding to the ordinary, heightened, and illumined states of consciousness. The author himself has traveled this path, and he knowingly sets up landmarks across the territory, guiding the reader toward deeper understanding. The quality of understanding gained through this training is what is needed to investigate in a meaningful way the basic issues of our existence, to question "why" and "for what purpose," and to regain a sense of wholeness and beauty of being.

In addition to such personal value, the TSK vision may contribute to a new methodology of knowledge, consistent with current developments in physics that aim at exploring subtler facets of reality. Let us remember the revolution in physics in the 1920's connected with the birth of quantum mechanics: the amazing discovery that light waves sometimes behave like particles, measurable in small portions (quanta); the even more astonishing finding that particles may display wave-like behavior; and the most striking discovery that atomic particles behave in an uncertain manner, so that not all their parameters can be simultaneously measured with unbounded precision.

All of this led physicists to believe that atomic phenomena do not fit our familiar logic with its fundamental principle of 'either-or', and even to invent a new kind of logic more suitable for quantum phenomena. This revolution implied the need to revise ordinary and previously accepted concepts of reality and to create new ways of understanding and reasoning and a new natural philosophy as well. Unfortunately, professional philosophers were taken unawares by these developments in physics, and until now have not been able to help create a new vision of reality consistent with quantum mechanics.

Now we are on the threshold of another breakthrough in understanding physical reality. Physicists have achieved a more or less successful description of the elementary particles of which our world consists, and they are trying to incorporate gravitational fields in the general scheme of things. This will have far-reaching consequences, since according to the modern view the gravitational field is synonymous with the metric structure of space-time. If these efforts are successful, we would be dealing with quanta of space-time, their creation and annihilation, and their transmutation into quanta of matter. Space-time would cease to be considered a container of things and a measure of events; it would become a living, fluctuating, constantly changing entity.

This new understanding will require, and perhaps is now requiring, revision of our concepts of space, time, and knowledge. We need an adequate philosophical framework and new training in mental perception to be ready for this new challenge. Hence the importance of the TSK vision. The approach it offers might help shape the required new view—an ontologically correct and existentially satisfying approach to understanding the outer world and acting in it in accord with our innermost essence, regaining a sense of wholeness and intimacy with our natural physical and mental environment that we still badly miss.

Space-Time And Creation

The TSK books raises questions that challenge all conventional views. For example, TSK students are asked to consider the nature of space 'in itself', as well as the relation of space to objects (TSK, pp. 9–10). Yet everyday experience suggests that space is empty and intangible, while objects seem wholly independent of space, which apparently serves just as their container. From this perspective, there is no access to TSK questions about space, nor is there any way to appreciate or make sense of such TSK insights as the key statement (TSK, p. 10): "Space is projecting space into space. That is all." [Editor's note: for more on this statement, see the article by Jack Petranker in *A New Way of Being* (Dharma Publishing, 2004)].

It might seem that the only way to explore such questions is by relying on subjective experience; in other words, by departing from the conventional wisdom regarding the nature of physical reality that we were taught in school. However, a closer investigation shows that is not so. Modern science gives some hints that might confirm the TSK concept of space from an objective point of view.

To see how this is so, it is necessary to turn from the ordinary world to the realm of the microscopic and macroscopic. In the domain of cosmology on the one hand and quantum mechanics and particle physics on the other, connections between objects and space manifest in new, unexpected ways, stimulating imagination and insight.

In the Newtonian conception of the universe, space is a container, absolute in nature: immovable, uniform, and homogeneous. Time is also absolute, flowing with mechanical uniformity irrespective of things and space. Objects are immersed in space and in time, which serve as background for the motions of celestial bodies, which are governed by the universal law of gravitation. This view had the virtues of formal consistency and physical accuracy, and so survived for more than two hundred years. In this century, however, it was superseded by Einstein's Theory of Relativity.

Einstein showed that space and time form a sort of medium that possesses a subtle but quite tangible reality. If we picture Newton's space-time as an absolutely smooth and rigid surface that simply supports objects but does not influence their motions and interactions, Einstein's space-time may be pictured as an elastic, pliable surface that takes active part in those motions and interactions. Every object 'lying' on this surface makes a depression in it that forces nearby particles to glide toward the object: this compelling tendency is the 'gravitational field' created by the object.

This interaction of objects with space and time is usually described as the 'curving' of space-time. If the object happens to move in an uneven manner, it creates ripples in all directions: 'gravitational waves' that can exist and propagate independent of any objects. Since space-time has its own sensed properties, it should be considered as having its own peculiar kind of reality. It interacts with the other kind of reality generally known as material bodies in a specific way: Matter shapes space-time, and space-time forces matter to move in a definite manner.

In Einstein's theory, space-time and matter are still independent, though interconnected, entities. This remaining barrier was broken by the discovery of the quantum nature of reality. Quantum theory predicts that all material entities exist in very small portions (quanta), and that quanta of interacting material entities can transmute one into another. There are quanta of matter (electrons, protons, and so forth), of light (photons), of sound (phonons), and of magnetism in solids (magnons). According to this theory, there also should be quanta of space-time (gravitons), though no such quanta have yet been discovered, and it is even difficult at present to imagine what such measured-out portions of space-time would look like.

Now, since space-time and ordinary matter interact, quantum theory suggests there should be transmutations of electrons, photons, etc. into gravitons, and vice-versa. That is, matter can disintegrate and generate space-time, and apparently 'empty' space-time may create objects.

These transmutative processes might be observed at very high densities of matter and in highly curved regions of space-time. Therefore, they should play an especially important role in the early moments of the universe, where exactly those conditions prevailed.

We are led to a question carefully considered in contemporary cosmology: the origin of the universe. Relativity theory coupled with particle physics predicts roughly the following stages in the evolution of the universe: (i) creation of the universe; (ii) a very short 'inflationary period' when the universe is spatially flat (uncurved) and homogeneous, being a very dense mixture of subatomic particles; (iii) the big-bang stage: an unimaginably intense explosion of the mixture; (iv) the period of forming ordinary material stuff: atoms, stars, and galaxies; (v) the current period of a very slow expansion of the universe, which may or may not later give place to its contraction.

While this theory faces various difficulties, perhaps the most challenging problem is the actual point of origin of the universe. Attempts to account for this event shed new light on the interrelation between space and objects. For instance, a current theory supposes that the inflationary universe emerged in a jump-like manner (as the result of a quantum fluctuation) out of an indescribable 'something', a state lacking ordinary matter, space, and time. Here indeed we have a KTS "zero-point of origin!"

We have thus reached a point where comparison between the scientific view and the TSK vision might be fruitful. For example, taking the cosmological view further, we might consider that at the originative point described above, the matter in the inflationary universe may have been represented by a homogeneous field, which at the end of the inflationary phase spontaneously transformed into a big-bang fireball consisting of material particles. In this view, it might accurately be said that "space projects objects into space," whereas space itself seems self-created.

Again, scientists attempting to specify this process of origination suggest that the dynamics of the inflationary universe can be likened to a ball rolling down a hill toward a new position, where it stabilizes due to a process of 'friction'. These occurrences—dynamic momentum, friction, and resulting stabilization in a frozen state or pattern—may be considered material counterparts of processes well-known to TSK students studying creation of the 'universe' centered on the self.

Are these theories and models relevant to the TSK vision? Nature likes to preserve its efforts, and it often reproduces the same patterns on different levels. This is reflected, for example, in the sameness of the mathematical laws that govern events at every level of patterning susceptible to mathematical description. Hence, it is not inherently improbable that processes operating in both the ordinary objective realm and in the subjective realm repeat the immense processes of creation that gave rise to the known universe.

Pondering the nature and function of space-time according to prevailing physical models may clarify some aspects of the TSK vision. It seems that all reality is permeated by connections and correspondences that help us unravel the mysteries of nature and human existence. "As above, so below; as below, so above." Studying "the inner life of material things" and the creative forces and processes of the objective realm may promote an appreciation of the workings of higher-level space and time.

At the same time, the TSK vision can immensely enrich scientists' stock of images and conceptions. The imagination sometimes fails to picture such strange things as "zero-point origin." The TSK vision may help stimulate thinking on a richer conceptual and imaginative level. Though precise at a deep level, the TSK approach supports the 'weird' and paradoxical style of thinking favored by theorists trying to uncover the mystery of creation. Perhaps not every such interplay will be fruitful, but it may also be that some such exchanges could lead to deep new insights that once more question structures we today take for granted.

Wholeness in Science

At the dawn of Western civilization, early Greek philosophers attempted to explain the world by reducing it to a single principle or a few underlying substances. For example, Thales of Miletus maintained that all things are composed of water, so that they may be viewed either as successive stages of water's transformation (ice, vapor) or as products of its densification (stones) or rarefaction (air). Early Greeks produced a number of such theories, all of which had a purely qualitative character. No exact predictions could be made, and they could not serve as a basis for appropriate action. Moreover, all of them contradicted one another, and there was no way to establish which one was true.

No wonder that a quite different line of Greek thought came to prevail, based on Euclid's geometry and Aristotle's logic. Aristotle's premise was that nature can be analyzed into parts and described by means of carefully formulated categories. This approach was at once enormously successful in descriptive sciences such as botany and physiology. Euclid considered bodies as consisting of partless parts (points) and bounded by sharply defined boundaries (lines or surfaces). This approach yielded an amazingly rich body of knowledge about geometrical properties of bodies.

Both these lines of thought were synthesized some two thousand years later by Newton. His approach made it possible to found such exact sciences as mechanics, optics, etc. and to establish the scientific method. By relying upon analysis and quantification, this method became very successful at analyzing substances into more and more minute parts: atoms and molecules, nuclear particles, and finally, the fundamental particles physics explores today.

However, in this process of progressive analysis, reality as a whole was lost. Attempts at reconstructing a whole reality from the scattered parts met with only limited success, so some scientists and philosophers went even further, returning to the ancient Greek concept of the world as a kind of organic, self-transforming, and multi-level entity. They considered the world an interconnected and indivisible wholeness, more fundamental than its parts (Lenikow, 1990).

According to David Bohm (1980), one of the foremost proponents of the philosophy of wholeness, the world can be viewed as a sort of hologram, encoding the whole in each of its parts. In his view, the holographic universe is essentially a combination of wave patterns associated with quantum waves carrying both information and energy. The human mind is able to read out those holographic patterns, and perceives them as images of the familiar four-dimensional world of space, time, and causality. The main feature of the universe is its whole movement, or 'holomovement'. Out of that holomovement certain spatial patterns may unfold that are seen by us as moving and interacting bodies, exactly in the way that waves on the sea surface can create shapes and forms.

Bohm considers space and time to be projections of a higher-dimensional reality. The following passage (1980, pp. 191–92) clarifies what he means:

> . . . [W]hat we perceive through the senses as empty space is actually the plenum, which is the ground for the existence of everything, including ourselves. The things that appear to our senses are derivative forms and their true meaning can be seen only when we consider the plenum, in which they are generated and sustained, and into which they must ultimately vanish. . . . This plenum is, however, no longer to be conceived through the idea of a simple material medium, such as an ether, which would be regarded as existing and moving only in a three-dimensional space. Rather, one is to begin with the holomovement, in which there is the immense 'sea' of energy. . . . This sea is to be understood in terms of a multidimensional implicate order . . . while the entire universe of matter as we generally observe it is to be treated as a comparatively small pattern of excitation.

> This excitation pattern is relatively autonomous and gives rise to approximately recurrent, stable and separable projections into a three-dimensional explicate order of manifestation, which is more or less equivalent to that of space as we commonly experience it.

Bohm has invented a special language of wholeness, 'rheomode', to describe the world as a flowing whole (from the Greek rheos, meaning 'flowing'). The necessity of such a language is based on the view that our language reflects our fragmented thought, intended to describe the world as a sum of fragments (particles of matter, parts of the human body, etc.), and that its structures condition our vision of reality. For example, the subject-verb-object structure of our sentences is a direct reflection of the subject/object dichotomy in our minds, which makes it difficult to express non-dualistic views using 'meaningful' speech. In Bohm's proposed new language, all words are variations of verbs, reflecting the holomovement as the nature of reality.

There are real points of contact between this new 'science of wholeness' (Briggs and Peat, 1984) and the TSK vision. For example, here is a passage (KTS, p. 84) that invites comparison to Bohm:

> Time's rhythm is transformed as it enters the realm of the 'outsider': It becomes 'points' and begins to build up a continuum. Understood in terms of external motion and the 'order' of the 'logos', rhythm builds from point to point of space and segment to segment of time. The waves of rhythm that originate with lower-level

time lead to origins, origins give boundaries, and boundaries call forth continuity to link together what is bound.

Still, there are important differences in approach. For example, whereas Bohm's approach is scientific, Tarthang Tulku's work is closer to art. He makes use of a multi-faceted approach in which the same subject is considered from various angles. These particular visions are like layers creating an image that in the end resembles a richly textured canvas. Even when the terms used are scientific in their surface meaning, this same 'impressionistic' approach applies.

Unfortunately, such an approach would be considered marginal by most scientists, so that the common points that do seem to emerge cannot serve to bridge the gap between the TSK vision and the established body of scientific knowledge. Other ways to bridge the gap must be found, e.g. in the area of scientific creativity or in the field of science education. Meanwhile, the parallels between TSK and the 'science of wholeness', a field becoming more and more influential, could be explored for the eventual benefit of both the TSK vision and science.

REFERENCES

Bohm, D, 1980. *Wholeness and the Implicate Order.* London: Routledge & Kegan Paul.

Briggs, J. and Peat, F.D., 1984. *Looking Glass Universe: The Emerging Science of Wholeness.* New York, Simon & Schuster.

Lenikow, A., 1990. *The Wholeness Principle: Dynamics of Unity Within Science, Religion and Society.* Wheaton, Indiana: Theosophical Publishing House.

The following books, all written by serious scientists, are also recommended:

Davies, P., 1988. *The Cosmic Blueprint: New Discoveries in Nature's Creative Ability to Order the Universe* New York: Simon & Schuster.

Greenstein, G., 1988. *Symbiotic Universe: Life and the Cosmos* New York: Quill/William Morrow.

Misner, C.W., Thorne, K.S., and Wheeler, J.A., 1973. *Gravitation.* New York: W.H. Freeman (a very reliable source of information concerning relativity theory, parts of which can be understood by lay readers).

Pagels, H.R., 1986. *Perfect Symmetry: The Search for the Beginning of Time.* Toronto: Bantam Books.

EXCAVATIONS

In *Knowledge of Time and Space*, Tarthang Tulku speaks of time 'excavating' space to give the structures of existence. In this section, we present various excavations, some directed toward the TSK vision, others toward experience and knowledge. In the end, these two forms of excavation meet.

The first article offers a gentle introduction to a different way of seeing and knowing. Javier Rockwell uses the metaphor of the labyrinth to describe our ordinary experience. Readers might want to compare this image to the many discussions in the TSK books of the stories and patterns that shape our lives. The metaphor of the labyrinth has two sides: one portrays it as a maze in which one loses oneself, perhaps forever; the other sees the labyrinth as a structured whole, almost a mandala, that leads inward toward a point of transformation. Perhaps there is a place where these two sides of the labyrinth meet as well.

The article that follows, on the future of time, is one of two I have contributed to this section (the other is on space). It is based on a talk presented at Nyingma Centrum Nederland in Amsterdam in October 2000. An earlier talk on the same theme was presented in August 2000 at the annual meeting of the Academy of Management in Toronto; for a paper related to that talk see http://www.creativeinquiry.org/fields-of-inquiry/Organizational%20Development/knowing-the-future. The article published here suggests that we could learn to *understand* time differently, and therefore to *be in* time differently. It lays out four different approaches to time that line up with concerns and approaches we take in living our lives: natural time, objective time, subjective time, and dynamic time.

Donald Beere's article illustrates nicely what the theme of excavation is all about. Reflecting on the TSK vision in light of his own experience, he is struck by the fact that the vision has little to say about awareness, intention, or choice, which strike him as central to his own explorations of TSK. And so he launches his own inquiry. To extend the metaphor of excavation, one might think here of a company exploring for oil in an area with proven reserves that decides to drill new wells in previously unexplored terrain.

In the short article I contribute on space, I question the idea that space is always empty, and suggest that there may be two different kinds of space worthy of exploration. This particular 'excavation' stays close to the themes of the TSK books, but turns the inquiry in a somewhat different direction.

Alan Malachowski writes of "paradoxes of transcendence" in TSK. He investigates two possible criticisms of TSK, each closely related to the other. The first is that TSK fails to clarify whether the instances of transcendence that it seems to promise (greater mental and physical health, vastly expanded knowledge, etc.) are to be understand as happening 'in reality', or only in a phenomenological or experiential sense. As long as this ambiguity persists, as long as the TSK reader is left in a state "suspended hermeneutic animation," the only readers for TSK are likely to be those who are willing to set aside the criteria for truth claims that have been painstakingly developed over the course of several centuries of inquiry in the West.

The second issue for Malinowski is whether transcendence is even a desirable goal. Tracing this question out, he is led to resolve the "hermeneutic suspension" that he identifies in favor of a phenomenological approach The question here is whether a TSK vision that becomes 'just another' approach to subjective transcendence is worth the effort. He closes by asking whether TSK can build a convincing case for autonomy from the naturalism that dominates virtually all contemporary philosophy As Malinowki suggests, these are questions that the next twenty-five years of TSK inquiry will no doubt continue to address.

It is fitting to close with an essay by Tarthang Tulku himself, one that builds on the most recent book in the TSK series, *Sacred Dimensions of Time and Space*, inviting new dimensions of research and new ways for the TSK vision to unfold. The author freely admits that the order he explores—based on the polarity of '0' and '16'—is completely arbitrary in terms of the structure of conventional logic, but sees in this no obstacle. He exhibits in this essay, as so often in the various TSK books, a remarkable ability to give experiential significance to what for any other writer would be wholly abstract—to find the 'feel' in the field.

These two qualities are worth considering in more depth. In the emerging history of inquiry guided by the TSK vision, we are at the outset, sailing on uncharted seas. If we are to discover new realms, we must be willing to lose sight of the familar landmarks and accustomed distinctions, to strike out into the open waters of the unknown. Since it within the vision of TSK that we are exploring, we must be ready to let go even of 'sea' and 'progress', of 'new' and 'uncharted'. The '16' with which this final essay closes can stand as a symbol for what we might discover in this way—not a place or substance, not a structure or outcome, but a fullness at one with openness, and a joy that emerges out of nothing at all.

A Magical Labyrinth

Javier Rockwell

To imagine a world in which reality depends on constructs does not mean retreating into fantasy. Instead, it means entering that world with its prevailing logic and its pre-supposed order as fully as we can. — Tarthang Tulku

As I write this, I am using a word-processing program that operates inside a well-known computer-generated 'environment' known as Windows. The Windows environment allows me to write and edit without having to know anything at all about how computers work.

In comparison to the low-level programs and routines that operate the computer, Windows is a fantasy space, superimposed on a more fundamental reality. By choosing to enter this closed fantasy space, I gain the convenience of being able to do my work without understanding the underlying technical machinations that make this process possible.

Yet living in a fantasy space also has its disadvantages. By giving up the opportunity to interface with the computer on a more fundamental level, I have given up a certain amount of control over my situation. All the technical, messy details that allow Windows to create the illusion of a user-friendly environment are hidden inside a 'black box'. What is inside the box is the privileged information of a few programmers. If anything happens that has not been anticipated by these programmers, then I become the helpless victim of their oversight, unable to leave behind the fantasy space of Windows and end the reality that underlies it. Dependent on the marvelous tools that let me do my work with greater ease, I am less flexible and less able to adapt to reality on its own terms.

Just as I inhabit the fantasy space of Windows when I use a computer, so I inhabit other fantasy spaces at other times. Fantasy spaces are created by music, mathematics, art, science, literature, and almost every other field of human creativity. Yet despite the great beauty of these rich fields, the result of living within these spaces is not what we would wish. By spending most of our time in fantasy spaces, we trade away the opportunity to develop our intelligence and flexibility for the short-term convenience or pleasure of living within an artificial and contrived simplicity.

Operating within fantasy spaces is not all bad. Provided we understand their power to confuse and disorient, we can actually use fantasy spaces to learn how to penetrate illusion. If we think of them as caricatures or distorted mirrorings of conventional reality, we have the opportunity to analyze and gently dissect them.

If we carry this process far enough, we find that our orientation changes. One day we may wake up in the 'real world' only to find this world—the realm of linear time and three-dimensional space—is itself another fantasy space; put differently, that conventional reality is itself a caricature and a simplification.

Why would this be desirable? Because our taking the conventional realm as real subjects us to continuous suffering. Like walking through a building still under construction, our efforts to traverse the conventional realm are full of unexpected obstacles and pitfalls. Unable to acknowledge what the conventional 'logos' does not encompass, we set ourselves up for confusion and pain. We see the results in our clumsiness and vulnerability, our frustration and emotionality.

For most of us, escaping from the limits of the 'conventional-reality fantasy space' will seem neither practical nor desirable. Confined to living within the structures of the mundane world, subject to all the artificial structures and fantasy spaces of modern society, we see no way to arrive at a more open view of reality that would not cause more pain and uncertainty.

What is needed is a way to cut through all levels of delusion at once. We need to see that reality somehow shines through all levels of obfuscation, no matter how twisted or confused. The challenge thus becomes: how does one take apart the conventional order in a way that is healing rather than destructive?

In looking at this question, I find it helpful to think of my life-experience as a voyage through an invisible maze or labyrinth. This labyrinth is the 'black box' that conceals the messy connections between conventional space and time and 'reality' as such. From a different perspective, it is also the magical bridge across the abyss that separates our artificially structured experience and the natural openness that underlies everything. The structure of ordinary time and space mirrors the walls of the labyrinth, but the labyrinth itself is more basic than that time and space.

When we accept the conventional realm as fundamental, we become aware of the labyrinth only through the unexpected consequences it brings to bear on our lives. For example, on a psychological level our response to the invisible barriers of the labyrinth might manifest as confusion or resentment. On an 'objective' level, the boundaries of the maze could express themselves as physical limitations such as the speed of light, the laws of growth and decay, the process of entropy, and other physical principles, known and unknown. On a mundane level, the walls of the labyrinth might show themselves as mistakes or accidents.

In general, we deal with the labyrinth by superimposing a fixed order—a fantasy realm—that describes interactions and relationships of entities and events. For example, on the physical plane there is the relationship of objects located in space as a function of time. To describe these operations, we make use of such concepts as distance and velocity. If we follow out these patterns of ordinary thought far enough, the conventional realm becomes frustratingly complex and nonintuitive—an indication of the labyrinth in operation.

Underlying this highly structured order is the experience of this order. Objects do not simply exist; they can be seen, felt, and thought about. The way they move and interact is associated with a capacity to sense movement and interaction.

The experience of the conventional realm is as rich and varied as the realm itself (if such a distinction can even be made). But a focus on experience offers a great advantage, because conventional experiences—the sights, the sounds, the smells, the tastes, the sensations, the feelings, the thoughts, and the emotions—can be ordered in a much simpler way than the objects of that experience. By focusing at this level, we begin to take the conventional order apart.

For example, all experience can be characterized as being denser or lighter, more tense or more open. We can explore these attributes, for instance, by looking for the lightest part of our experience and the heaviest. Once we know where these are, we can work to create a bridge between them. By forging connections between the light and stable aspects of our experience and those parts which are painful and obscured, we begin to find a way to balance our being.

From my own experiments so far, I can say that this way of working with experience cuts across the ordinary conceptual boundaries that separate us from the labyrinth. When we tune in to the felt sense of experience rather than a fixed and objective 'reality', everything can remain much more fluid. We have the chance to engage the labyrinth directly. instead of being bound within its confines. We have the opportunity to learn in a way that is both healing and meaningful.

Living at the Edge

Jack Petranker

For many years now, the idea has been current that we would enjoy ourselves more and benefit more from being alive if we learned to live in the moment, to "be here now." Of course, since TSK likes to ask questions, we cannot stop there. As soon as we say, "live in the moment," we want to ask, "What's involved in 'living'? And what do we mean by 'the moment'? And for that matter, are we completely clear about 'in'?" We could put the same questions about 'be' and 'here' and 'now'.

These are not just idle questions, the kind we might sit around and discuss in a café. We could do that, of course, but it seems more fruitful to acknowledge that these are fundamental questions about our nature as human beings. They invite us to reflect on what we can do to live our lives in each moment in the fullest and richest and most satisfying ways.

TSK gives us the opportunity to ask these questions at the most fundamental level. When we involve ourselves in the study or practice of TSK, we are raising the question of how to engage the 'nowness' of now, the 'hereness' of here and the 'being' of being. These are not questions that have answers, but they are definitely very interesting to explore. Each time we embark on such an exploration, we start over. Each time we must find a new way to bring our questions alive.

In this article, I will approach these questions by first developing a conceptual structure. This is unusual for TSK, which is not about theory at all. But structures can be pointers. Without structures, without concepts, our explorations might well start off blind and end up fruitless. In any case, the structure I will lay out is meant to serve as a tool. If it helps open the door to a different kind of inquiry, it will have done its duty, and can be discarded. As in the TSK books themselves, what is said here is not meant to be 'the truth'.

The focus of the structure is time. I will be presenting four approaches to time, which I will call Natural Time, Objective Time, Subjective Time, and Dynamic Time. What I am calling "Dynamic Time" is closest to the TSK perspective. If the build-up to Dynamic Time through the three other kinds of time seems too theoretical or abstract, please be patient.

Natural Time

Natural Time is the kind of time that human beings seem to have operated with in most cultures, including ours, until the last few centuries (perhaps until the clock began to exert such a strong hold on human consciousness). There are aspects of Natural Time that could justify calling it Sacred Time, or perhaps Cosmic Time, and from another perspective Natural Time could also be called Embodied Time. I call it Natural Time in the present context for the simple reason that it is the understanding of time that people seem to fall into naturally as they live out the full range of human experience. It includes embodied experience—the way we engage the world through our bodies and senses—and it also includes the parts of our experience that make it meaningful.

Natural Time relates to a sense of a flow within the cosmos.There is a sense that human beings are in harmony with the universe, that events are unfolding in a way that has meaning and in which human beings have a role to play. The task of human beings is to attune themselves to that flow or dynamic, and to act in harmony with it. Anyone who falls out of harmony must find ways (for instance, through ritual, prayer, and offerings, because usually harmony is linked to a sense of the divine) to restore balance and order.

In our culture, the feeling is widespread that we once lived in Natural Time and no longer do, and that we have lost something very important as a result. There are those who say this is just the price of maturing as a species—letting go of myth and illusion—but most people would like to get back in touch with that way of life and that quality of time. The hunger to do so may well be the motivating force that inspires many individuals to undertake spiritual journeys. Whether we have personal experience with Natural Time or not, we accept that we might be able to live in harmony with the flow of the universe.

In the aspect of Natural Time that I referred to above as Embodied Time, issues of meaning do not arise in the same way. In Embodied Time, we are part of physical nature, and the harmony we feel is physical, 'gut-level harmony. Examples might include the feeling people get when they dance, or play a sport, or make love. Still, there is the same sense of a natural flow to events, a flow in which we participate.

In terms of the fourfold structure I am working out in this article, the sense of loss and disconnection from natural time, the sense of having drifted into a different and less satisfying way of relating to time, evolves in this way: at some point Natural Time splits in two. After the split, there two times in operation: Objective Time on the one hand and Subjective Time on the other.

Objective Time

Objective Time is the time that scientists are concerned with, the time that we use to measure things out. Natural Time has measurement too—cycles, years and seasons, dynastic eras, and so on—but in Objective Time measurement becomes central. A grid is laid out (hours, minutes, seconds, and so on, down to nanoseconds). Once this trend toward measuring time out is set in motion, the units of measurement tend to get more precise all the time. This precision is important, because by measuring with precision, we learn to 'make sense' of nature and to control it. In contrast, the sense of control in Natural Time is much weaker. To simplify, in Natural Time our job is to align ourselves with the forces that shape the cosmos, or perhaps influence them in some way, but not to control them. This approach makes sense when there are meaningful structures in the universe. In Objective Time, however, meaning is no longer relevant. Time is just the tick . . . tick . . . tick of the clock. It is completely machine-like. So we deal with it as we would deal with any machine, learning how to operate it and put it to use.

When we are engaged with Objective Time, when that is what interests us, our focus shifts to 'figuring out', explaining, and making schedules and plans. That is the way of thinking that goes with Objective Time. First 'this' happens, then 'that': one event follows the next. As Henry Ford, inventor of the assembly line, famously remarked about history, "It's just one damn thing after another." Relying on Objective Time, we analyze, we predict, and in this way we control.

One way to think about the difference between Natural Time and Objective Time is that in Natural Time it makes sense to speak of "natural law." Certain things are 'right', in the sense that they fit in with the way the universe works. In contrast, in Objective Time we talk about "laws of nature." This sounds very similar, but it has a completely different flavor to it, because 'nature' understood in terms of Objective Time is like a machine. If "natural law" has to do with what is right, "laws of nature" have to do with the way things work. Instead of questions about what is right or wrong, laws of nature deal with what is true or false, with 'is' and 'is not'. This is a profound shift. It signals a loss of trust in the cosmos as meaningful, a loss epitomized in our transformed understanding of time.

The same shift could also be described in more positive terms. In the past we thought that nature was beyond our power to control, but gradually we realized that we could take charge of our own destiny, or at least improve our lot.

Positive or negative, though, it comes down to the same thing: we decided that we had better take matters into our own hands, and that measuring out time let us do this. The outcome is that time becomes the "tick . . . tick . . . tick" of one moment following the next—what TSK calls linear time.

Objective Time and the Past

If our interest is in controlling nature, our focus will be on the future, on predicting and affecting what is going to happen and getting desired results. In terms of knowledge, however, the same approach actually requires us to focus on the past. Objective Time is intimately linked to measurement and to collecting and manipulating data, and data is always based on the past. When we measure or gather information, we are looking at what has already happened. Even if we collect data in 'real time', the information we get reflects a past state, even if the gap that separates the past from the present is as small as a nanosecond. So although we may be predicting or aiming at a result, we are guided by what has already happened. It is a lot like steering a car by looking in the rear-view mirror.

Of course, if you drive by looking at the rear-view mirror, you are almost certain to crash as soon as the road up ahead changes from the road behind. And that in fact seems to be our present experience, because the road we are driving along is also a road we are building as we drive. The more we rely on Objective Time and the kind of knowledge that goes with it, the more we seem to construct a road that twists and turns in unexpected ways (what is sometimes called "the law of unintended consequences"). We drive along, gazing intently in the rear view mirror, with no idea what is coming around the next curve, or even when a curve is coming. Since we are so good at measuring time in very small increments, we can adjust for our mistakes, like a driver constantly making corrections to stay on a steady course. But fundamentally we are driving blind, always about to crash. It is a very hard way to live. No wonder so many people yearn to return to Natural Time, where the road is straight and level, and where knowledge is not so much past-oriented as timeless.

Subjective Time

I said above that in Natural Time the world is naturally meaningful, but in Objective Time meaning disappears. When we carve experience up into ever small units, there is no room for meaning. Yet human beings are meaning-giving and meaning-seeking creatures. So where does meaning go? Put differently, where can we go to find meaning? The answer is that we have learned to find meaning in the domain of Subjective Time.

Subjective Time is the time that is formed by the stories that we tell and the stories we live. (This distinction—between telling a story and living a story—is important, but for present purposes I will treat these two ways of relating to stories as equivalent.) Subjective time is story-telling time. It comes into play whenever we say, "This is the way things are," or "This is what's happening." Since it is always possible to tell a different story, there are countless kinds of subjective time. This is quite different from Objective Time, of course (or from the *story* of Objective Time!), because there is only one Objective Time.

Now, it would be possible to tell the story that says the only real time is Objective Time, that Subjective Time in all its variations is just an elaborate fantasy. Those who think this way say that if we human beings would only be realistic about the nature of reality, we could live a much better life, because we would be more rational and more effective, without messy emotions to worry about or conflicting stories to fight wars over. This *view* enjoys growing support, but most people cannot *live* in accord with it, and if they gave it a try would not even want to.

Instead, most people live with the sense of meaning in their lives: not necessarily the meaning that emerges in Natural Time or a Natural Order, but meaning that is personally significant nonetheless. In fact, this sense of what is personally meaningfully is so significant that people often seem to spend their lives wishing they could align their experience and actions with what they find most meaningful and bemoaning the fact that they cannot do so. At this level, Subjective Time approaches Natural Time, and we may confuse one with the other. The distinction is this: in Natural Time, the whole of the cosmos tells the same story, and that story extends beyond us as individuals to embrace all living beings. In Subjective Time, the story is one we are telling; it is our story, based on 'my' subjective reality: 'my' wishes, 'my' projects, and so on. Once Subjective Time is dominant, if we do tell the story of a natural order and of meaning in the cosmos, there is always the suspicion that it is just 'my' story.

Compared to Natural Time, Subjective Time offers a great capacity to adapt to the unexpected, because we are always free to tell a new story. In this sense, the ability to live in Subjective Time is just as much a part of the successes of modern culture as the the ability to measure out Objective Time. Still, Subjective Time is not as flexible as we might think, because our ability to tell new stories is constrained by our commitment to the old stories—our commitment, in other words, to the past. For instance, we are committed to our own identity, and we will tell no stories that reject the story of that identity.

Like Objective Time, Subjective Time is focused on the past by its very structure. When we tell (or live) a story, in a certain sense we are making it up as we go along. But in another sense, a story only makes sense in a backward-looking way. Stories aim at the past. Their structure is always the same: "Then this happened." For stories that we tell, we can only start the telling when everything is done and over with, as in the time-honored fairy-tale opening: "Once upon a time there was a king." Even if we are living the story in 'real time', we are implicitly making sense of it retrospectively.

This is perhaps most clear in dreams. For instance, suppose I dream I am in Amsterdam, but then suddenly find that I am in Paris. Something fundamental has changed. My immediate response is to tell a new story that makes the new setting makes sense. To do this I have to retell the past that got me to this present moment. The new story does not have to make complete sense; it just needs to hold together enough to let me keep going, with perhaps some lingering sense that something is strange. Looking backward, the pieces have to fit together. That is what I mean by the backward focus of Subjective Time.

Healing the Split

If we focus on Subjective Time, it begins to seem that nothing is true, since everything is just a story. "Everyone is entitled to her own opinion," as we say. From this perspective, tolerance is the great virtue. But this kind of relativism goes only so far, because most of us accept not only Subjective Time but also Objective Time, and in the Objective Order things are very definitely true and false. We live with a kind of conflict: in one part of our lives Objective Time counts, and in another part of our lives Subjective Time counts.

This division can become a major concern, because in our society Objective Time 'colonizes' the subjective world. For instance, consider what it is like to make a phone call to a big business nowadays. Very seldom do we immediately get to speak to a real human being. Instead, we have to go through a menu of choices, pushing the right buttons or saying the right words. This is Objective Time at work, classifying and dividing in the name of efficiency, an essential Objective Time virtue. Subjective Time—the time of human action and interaction—is pushed off to the margins. And even once it has been squeezed to the side, it continues to be colonized. For instance, think of how advertising invites us to construct particular kinds of subjective stories that can be fulfilled only by buying the advertised product. This is Objective Time taking over—not so much out of some evil intention, but because Objective Time inherently tends to put impersonal forces (in this case, economic forces) in control.

Perhaps this is one reason we currently see so much energy going into ways to expand Subjective Time. Think of virtual realities, and of the popular theme that our everyday reality is a fraud or a pale imitation. (I am thinking here of the *Matrix* movies, but also the *Harry Potter* books.) The illusory nature of everyday reality is a commonplace in children's literature, perhaps because for children framing their own subjective reality is an all-consuming concern. But now it seems to have made its way into the adult mainstream. Of course, in movies and books, the Subjective Order into which we escape is still somehow shared by others. That's what makes them fantasies.

The desire to heal the split between Objective Time and Subjective Time also manifests in wanting to return to Natural Time, to a 'timeless' world of wonder and enchantment, 'before' subjective stories or before the rigid order of linear time. One aspect of this is the nostalgia for a simpler way of life, simpler both in the sense of being less complicated and in the sense of being shaped by a single, universally shared set of values or a single understanding of how the world makes sense. Another aspect, more related to Embodied Time, is a return to the sensuous, to immediate experience and the body. In both cases, the sense is that we want to restore to our lives a kind of aliveness as well as harmony. So we aim for less complication, for release or relaxation and a feeling of harmony or oneness.

In social terms, we see this longing for Natural Time in various kinds of fundamentalism on the one hand and different forms of spiritual practice on the other. This would be too complicated to go into here. Instead, I will focus on experiences related to Embodied Time, with which we are all familiar. The hallmark of these experiences seems to be a kind of aliveness, as though the energy of time were coursing through our veins. A good example is risk-taking: skydiving, river-rafting, love affairs, and so on. On a more everyday level, people go to a horror movie, or go dancing, or take a vacation to some exotic locale. Or they may seek out simpler ways of living, more in harmony with the earth, or practice martial arts and yoga (which of cours also bring into play the spiritual dimension), or run in marathons. If you ask them why, they will often say, "It makes me feel really alive."

There are also other ways to heal the split. For instance, there is intense aesthetic experience. Or there is the kind of experience we may have in meditation, when first the rational, planning mind drops away, and then the story-telling mind drops away, leaving something closer to pure experience, energized but also calm.

Why do these various experiences make us feel intensely alive? You could say that they engage our senses deeply, or wake us up in unusual ways, or let us experience very strongly our embodied being. In terms of the structure being developed here, I would say it is that either they activate a link to Natural Time, or they mimic the harmony of Natural Time by recreating it at the subjective, story-telling level. Seen in this way, you might even say that the wish to unite or restore Natural Time operates across the whole range of human desires. Of course, acting on this wish can happen at lots of different levels. An escape into fantasy is not the same as connecting with embodied experience, and the joy of dancing is different from the joy of a profound meditative or religious experience. But the similarity is there as well.

What Remains Missing

As a way of making the bridge to Dynamic Time, I want to suggest that for all their positive aspects, such experiences—even the most profound—leave something out of account; put differently, they are inherently limited. (Readers familiar with the TSK books might compare the difference between second-level time or knowledge and third-level time or knowledge. See especially chapter 29 in *Love of Knowledge*.)

The first important limitation is that if we make it our goal to seek out such experiences, then to feel fully alive, or at least more alive than usual, we will have to put ourselves in a special situation or do something out of the ordinary. Assume we succeed; nonetheless, the rest of the time, when those special circumstances do not come into play, we will remain stuck in our subjective stories, or in someone else's stories, or in the Objective Order. It is sad to think that this is the best we can do, because it means we are living for the sake of a few special moments or special kinds of experiences. The rest of our lives becomes 'down time', during which we can only crave more of the experiences that make us feel truly alive, or daydream about them, or talk ourselves into believing they are happening.

Dismissing the value of much of our lives in this way is the first limitation on touching the aliveness of time by aligning with the rhythms of Natural Time. The second limitation is more subtle. It has to do with an inherent tension between the immediacy of these kinds of experiences and our ability to deepen our knowledge or self-understanding. This is a complex topic that would require a more careful analysis than I am prepared to give here, but I do want to at least suggest the basis for my concern.

If we are most alive in extreme situations—extreme joy, extreme fear, extreme love, extreme novelty, and so on—our ability to deepen our understanding and knowledge will be correspondingly restricted What makes these situations attractive to us is that they involve us completely. But when we are involved completely, do we 'know' what is happening? If the experience absorbs us completely, how can knowledge arise *within* the experience? I do not mean to suggest that knowledge requires that we distance ourselves from our experience (that would be a view more in keeping with Objective Time). Still, to be absorbed completely is in some sense to be lost. Does this 'being lost' mean we also lose the chance to deepen our understanding?

For instance, if I taste something wonderful, I may 'know' that taste very richly. But what about knowing *myself* in the act of tasting, knowing my reactions at various levels and in various dimensions of experience? What about being able to entertain the possibility of reacting differently?

There is a sense in which knowledge, even profound knowledge, seems too 'cool', too uninvolved, to be able to flourish in the intensity of a Natural Time experience. To see this, consider what experience is like for animals. It seems reasonable to say that animals live in Natural Time almost exclusively. Then where does access to knowledge come in? Or think about small children. The immediacy of their experience seems almost admirable, or at least enviable. But what about knowledge? If we are interested in changing who we are, in becoming more than we now are, does the absorbtion of Natural Time offer that possibility?

I said earlier that Objective Time and Subjective Time were both centered on the past, and that from a Subjective Time perspective this focus leads to a kind of nostalgia for what is understood as the 'timeless present' of Natural Time. Suppose we learned to live in that timeless present, a 'standing now', lost in the immediacy of experience like so many 'noble savages'. Is that truly the best option we could pursue? My sense is that something would be missing, a dimension of greater knowledge—what TSK calls 'knowingness'. And this leads me at last to Dynamic Time.

Dynamic Time

In *Dynamics of Time and Space,* Tarthang Tulku writes, "Through time, the universe comes alive." That aliveness is Dynamic Time. It is just what Subjective and Objective Time are lacking. When we tell a story about events in time, or when we measure out time (i.e., 'tell' a story as clocks 'tell' time) we do not touch time's dynamic aliveness. The story and the telling keep time's dynamic at a distance.

In contrast, in Dynamic Time, the aliveness of time is available *in every moment.* This "every moment" way of talking may seem to take us back to Objective Time, but that is not so: measuring moments in a linear way does not use them up. When we let moments slice time up, its aliveness eludes us, just as the aliveness of a human being eludes someone dissecting a corpse. But there is also the *passage* from moment to moment. That is where the dynamic operates.

How do we engage that dynamic aliveness? The TSK books offer countless leads and many exercises toward that end, but let us look anew. It clearly will not work to take an Objective Time approach, regarding time as a kind of 'thing' available for us to take hold of and examine until we find the part we can label 'alive'. That way of looking demands that we freeze time, stop it in some way so that we can investigate it. It kills the aliveness for the sake of the investigation. On the other hand, we will do no better if we take a Subjective Time approach, for then we are simply telling a story about time, or perhaps telling a story about our stories.

If we try to stay independent of both Subjective and Objective Time, if we stay true to what is natural in Natural Time, we can say that the access point to the aliveness of time has to be in in our experience in this moment, right now. How do we enter this 'in' without immediately emerging on the other side? How do we dwell within the 'in' of time?

Dynamic Time and the Future

Given the structure of this article, the answer I give to the question just posed should be easy to anticipate. If Objective and Subjective Time commit us to living in the past (and thus living without access to time's dynamic), and if Natural Time immerses us in a timeless present where something essential to knowing is lost, Dynamic Time invites us to live in the future.

The future has no content, no stories, no structure—it is completely open. The future is unknown, because there is nothing there to know.

Compare this to Natural Time. Experiences of Natural Time are alive, but they also insist on a certain structure or identity (down to the 'pure experience' identity of "nothing at all.") This is a way of restating the limitation on Natural Time experiences that I discussed above: in such experiences, we lose ourselves in the identity of the experience. In contrast, the aliveness of Dynamic Time has neither identity nor content.

In fact, it may be that wherever we encounter aliveness (for instance, in the flow of Natural Time, but perhaps also in the *creativity* of Subjective Time or the radical *deconstructive thrust* of Objective Time), we have entered Dynamic Time. In that sense, Dynamic Time is not a separate dimension of time; it is the aliveness of time in all its manifestations.

Linking Dynamic Time with the future may seem to leave to question of knowledge problematic. I have suggested that the timeless present of Natural Time gives no access to knowledge, but where is the knowledge in the open future of Dynamic Time? How can there even *be* knowledge without structure or content? The answer is that Dynamic Time gives access to a different kind of knowledge. 'In the future', we can know the dynamic of time itself. We know it by inhabiting it, by living (in) it.

Because living in the future frees us from all past-centered limits, it frees us as well from all limits on knowledge. This way of engaging time does not give us greater access to the kind of limited knowledge to which we are accustomed, does not expand or extend those limits. Instead, it gives us access to unlimited knowledge—not more of the same, but a different kind of knowledge completely, a knowledge not bound by past-centered content and identity.

How can we live in the future? We cannot just pack our bags and take a train that deposits us in the future, but fortunately that is not required. The future is already 'here', right at the edge of our present experience. We live in the future when we live 'on the edge'.

Given a flow of time centered on the present, the edge seems to make itself available automatically. In order for anything to happen at all, the future has to be arriving. If time was not already active and dynamic, none of 'this'—'the way things are' and 'what's happening'—would be unfolding. The very possibility for presence and experience would not manifest. That possibility is not some remote abstraction: it is the truth of our being. And it is available right now, at the edge of what's happening.

The possibility for presence and experience never takes form, even though it appears together with form and involves content. Always at the edge, always arriving, it is presented by the future. Unlike the content of experience, based on the stories or data from the past, the actual experiencing—the 'something going on'—is not 'based' on anything. It is just the arriving of the future, open for any kind of content. The future can be our being—the 'when' of what we are—at the same time that it is nothing at all.

There is no mystery about this, even though trying to put it into words may make it sound mysterious.

The dynamic of time is always available, and we can engage it right now. Go ahead and try. . . now . . . and now . . . and now. Do you get a sense? You can keep on engaging while you are reading, or even as you wonder off into a daydream.

That is what I mean by living at the edge. *Dynamics of Time and Space* (p. 99) calls it the "point of arising," the place where the future "comes into being." Can you get a sense of that right now? If you can, consider that you do not ever have to stop engaging the future in this way, because it is always available.

Of course, you *will* stop. Why should that be? Why can I make that prediction with certainty? Why do we disengage from the future?

The answer is that living at the edge, or living in the future, seems to run counter to the basic structure of the self. The self is built up on the basis of identities and stories, and so it insists that something is static, that not everything remain in flux. Because it turns away from content, living at the edge is inherently playful and spontaneous. But the self is the guardian of content, and has no wish to play, to dwell on the edge.

To counter the gravity of the self, we to draw the self into our practice, into our play and our experiments with time. It will not be adequate to say, "I'm going to do this exercise and see what kind of experience I have." In that way of talking and thinking, we leave the self out of the picture, and that amounts to tricking ourselves. To leave the self *out* is to leave the self *in place,* and this in turn freezes time's dynamic.

Precisely because it goes against the patterns of the self, engaging Dynamic Time is a challenge. Who does it? How can the 'who' that is doing it challenge itself? Fortunately, the problem sounds more real than it is. There is no hard work involved, no convoluted conceptual structures to overcome. The river flows with no one pushing it. The question is: Do we float in the river, or do we become the river? As soon as we get out of the way (into the way?), Dynamic Time makes itself available. It is not some unusual event, not a party to which we have to receive an invitation. It is just what is always happening.

One more time, let us compare this to Natural Time, because from our subjective/objective perspective it is easy to confuse the two. Natural Time has limits, because in natural time there is always a 'way things are,' a past that shapes the 'what's happening' of the present. With Dynamic Time, there are no such binding structures. Of course, structures are always taking form; they are always on the edge of being established. But we live on that same edge, in the arriving future. We are always at the beginning, or even before the beginning, in a realm of unending possibility.

What can you do to activate this possibility? Just remind yourself! Be ready to look in places we do not usually think to look, and to ask in ways we do not usually think to ask. Then Dynamic Time inquiry can be a lot of fun. It is like the 'fun' of feeling really alive (Natural Time again), but it is more than that, because it sets no limits. It just opens up into total possibility, and that is the most fun of all.

Awareness, Intention, and Choice

Don Beere

Throughout the many books Tarthang Tulku has written on Time, Space and Knowledge there is almost no discussion of awareness, intention and choice, yet the exercises that infuse these books require all three. The lack of overt references to concepts that implicitly pervade the texts is at odds with the sharp and incisive critique that characterizes most of these works. This discrepancy has nagged me for years, and I want to use this article as an opportunity to explore the incongruity, specifically in regard to doing TSK exercises. As I have done previously (Beere, 1980; 1997), I take a phenomenological approach. The focus is on what phenomenology calls the "natural attitude;" what TSK would call level-one experience. As pointed out in the TSK books, level-one experience is always potentially transparent to level-three experience. It should be possible to see 'through' levels, because ultimately the three levels described in TSK are one.

The definitive intention of this essay is to deepen understanding of TSK and to further its practice. A more immediate intention is to explore how choice can alter the subject that chooses as well as the very structure of experienced consciousness on the one hand, and the structure of experience on the other.

The Relation Among Intention, Choice, Action, and Will

I will start with an everyday example. Let us say I have the intention of crossing a stream. There is a reason to cross. For example, I want something on the other side—a better view, flowers to pick, or safety. I have yet to move, since I have only formed the *intention*. To get to the other side I must move, and that involves the *decision* to act followed by the *choice* to move followed by *specific action* to make actual the intention, as in the following figure:

Plan		Initiation of Plan		Action Choice		Goal
Intention to reach goal	•	Decision to act so as to attain goal	•	Sequence of choice of action followed by action	•	Reaching the goal

The Process of Reaching a Goal

How I act involves a series of choices followed by actions: choices made at each step. Once I have formed the intention I must still select what I will do. I could go directly across the stream, but I could also take any other route, no matter how indirect. Again, having the original intention or plan, I still might not decide to initiate that plan. For instance, I might delay beginning to cross until I find an easy opening in the underbrush. I might, having started, give up. Movement toward the goal requires an ongoing series of intention-choice-action sequences. Completing an initial series of actions does not guarantee I will complete the next until I eventually cross the stream.

It might strike the reader that the distinction between choosing to do something and choosing to act on that choice makes no sense, since it makes the first 'choosing to do' a nullity. For instance, suppose a high-school student chooses to go to college but never sends in an application. Wouldn't the best description be that the student never chose to go to college at all, but only pretended (to self or others) to have made that choice? Is a choice that does not lead to action really a choice?

I would respond that I use the word "choice" in a different way. Using my terminology, the high school student has formed the intention of going to college and seriously plans to do so, making this public by sharing it with others. To achieve that goal, at some point the student must engage in a number of actions, beginning with the application process, which itself has many steps and options, starting with the decision as to which colleges should be considered. The student might genuinely plan to go to college but never decide to do so, or decide too late, or discover further down the road that the application process is overwhelming.

Another everyday example is the plan to go on a diet. Once the plan is formed, the decision to begin might be postponed to after the first of the year, or in time for summer. Next the individual must choose what actions to take to accomplish the plan; for example, restrict calories, join a commercial weight-loss program, or eliminate certain foods. To realize these plans requires a series of choice-action sequences over weeks if not months. Frequently, challenges arise to sticking with the plan after its initiation: tasty and forbidden food invites consumption or hunger becomes difficult to tolerate. The plan must be maintained and reaffirmed at every step.

A More Detailed Investigation

In the diagram presented above, the words used are nouns, static and final. Experientially, however, aware ness, intention, and choice involve a process, which suggests verbs: action-oriented words. With this in mind, let us now look more closely at the terms.

Intention According to the *Random House Dictionary of the English Language,* (2nd Edition), "intention" is linked to the verb "intend," whose Latin root means "to stretch towards or aim at." The dictionary offers as one definition "to have in mind as something to be done or brought about; plan." Intention, let us say, is the state of having a mental aim or direction, preparatory to decision, choice, and action.

Goal A goal can be as lofty as "enlightenment" or as mundane as relieving the discomfort of an itch. The goal can be regarded as what the intention directs, aims at, or "stretches toward."

Decision A decision links to the verb "decide," defined in the dictionary as "to make up one's mind as to what shall be done and the way to do it."

Choice links to the verb "choose," which can mean "a decision on one of a number of possibilities because of its apparent superiority." Decision and choice are similar, but in the context of this paper, decision refers to the mental state associated with a commitment to achieve the goal, whereas choice refers to selecting a specific action from a number of possibilities.

Returning to the diagram on p. 233, I can have an intention without deciding to achieve it yet or without acting. I might have the intention to quench my thirst by getting a drink of water, but not yet: "I'll get that drink after I finish this paragraph." An intention is the mental 'pointing toward' a goal; to reach the goal, I must decide to act on the intention. Each action is chosen from a variety of options and sculpts how I reach the goal.

In the example of crossing the stream, as I proceed toward the stream, I might encounter obstacles. I might get tired, the weather might change, the stream might be too swift to ford, the footing might be slippery or marshy, or I might discover rocks I cannot negotiate. I might give up on my original intention. To continue requires *will.* I must maintain my intention in awareness, I must continue making choices about how to move toward it, and I must act on those choices in the face of inevitable obstacles and possible discomfort.

To achieve any goal requires experiencing it as "over there" or "not yet" and engaging in subject-initiated action to attain it. This structure demonstrates the subject's uniqueness in the world. I am in some way distinct from and can influence the world around me; vice-versa, the world is distinct from me and can influence me. This is implied in one of the few passages in which Tarthang Tulku refers to the intentional experience of the subject (volition):

> Time is then handy; it expresses the distance between the self and its object, thus delineating and ensuring their difference (according to the self's view). The power of Time fragments, becoming the volition of the self, on the one hand, set in contrast to the energy of nature and physical processes, on the other (TSK, p. 169).

Application to TSK Practices

Most of the TSK exercises involve mental activity of one kind or another, whether observing experience, exploring awareness, or generating mental phenomena such as images. A struggle like that of crossing the stream can arise during these practices. Consider the Giant Body Exercise (TSK Ex. 1, p. 21) in which the reader is asked to create the image of a living, giant human body in all its details. I must first have the intention-decision followed by the choice to engage in creating this image. Next I engage in mental acts that bring into being various elements of the image. I begin, let us say, with the foot, and discover that I can generate clearly one but not two toes. I can create a live, wiggling toe, that has hair, "real" wrinkles at the joint and even realistic texture to the skin. These details disappear when I add a second toe. I keep trying with the same disappointing result. I begin to get frustrated, and then discover myself tired, a tiredness that seems associated with the effort of constructing the image. To succeed I must continue despite the difficulties and obstacles. To continue requires will.

None of this is specific to TSK. Anyone who has meditated has confronted these same experiences. If the meditative practice is following the breath, I generate the intention to follow the breath, followed by the decision to act on that intent. I then perform the act of focusing awareness on the breath. As the meditation continues, the mind might shift away from the process of breathing; for instance, noticing an ache. Thoughts about the achy, bodily region and then about the body in general appear in the mind, followed by images of activities linked to that ache. At that point I may become aware that I have shifted away from the breath and, as a result, bring my attention back to the breath. "Bringing my awareness back," a mental action, requires the same sequence as crossing the stream: intention-decision-choice-action. When obstacles arise —when I get tired, distressed, bored, lose motivation, or find other interests arising—I must use will to keep awareness on the breath. As those who have meditated for prolonged periods know, the effort becomes both mental and physical.

All of the TSK exercises entail the intention-choice-action sequence, maintained over time via will. Having practiced many TSK exercises a great deal over the past 20 years, these are processes I have struggled with frequently. I have also used the TSK books as texts in both graduate and undergraduate classes, and my students report similar struggles, invariably remarking on how difficult and demanding the practices are. This is one reason why I find it puzzling that a discussion of choice and will is notably absent from the TSK books.

The TSK exercises involve awareness and attention, also not discussed extensively. The following section explores how awareness and attention relate to choice and will. In tracking their presenting, it will become apparent how, moment to moment, we make choices based on intentions.

Awareness and Attention

Awareness can be likened to a pervasive field that surrounds and pervades the experiencer or subject. On the one hand, awareness is so encompassing that it has no specifiable meaning. On the other, it is so pervasive that it lies at the heart of what we experience and what we know. *Attention* brings into focus various aspects of what is or could be in awareness. Awareness continues; attention focuses. Like a spotlight, attention makes central its object. In some ways, awareness and experience are the same; in other ways, attention guides experience, brings it into being, and shifts it from this to that. One might describe this in the following way: experience is what attention takes away from awareness. We can clarify this through an example:

I sit at my computer struggling to find the words to capture 'awareness'. I am vaguely aware that my fingers are on the keyboard and I see but do not read the words on the monitor; my awareness is primarily focused 'inside' on my thoughts and the words I will type to 'capture' awareness. The thought occurs: "Each time I capture awareness, it slips away." I smile at the irony. My task is more like pointing than writing a definitive exposition, and wherever and however I point, awareness is there and here and in between.

The focus of awareness shifts back to "sitting at the computer in the study." I am aware that my immediately preceding awareness was more interior, on thoughts and ideas—some thoughts articulate, others diffuse. I was, however, not unaware of sitting in the study: that aspect of awareness seemed less central, almost as if dimmed. It seemed to become distant as the focus centered on thoughts. I realize, for example, that I was distantly aware of my wife typing at her own computer. As I reflect on this, awareness is filled by the room, her presence, and the sound of her typing, and the inner contents and operations of my mind fade. I recall (notice in retrospect) that when I was aware of my thoughts, my mental interior, it was diffusely associated with my physical being, my body, the sounds in the room, and my presence as a subject, as the observer, thinker and experiencer. What is in focus, what is centrally in attention, is what I would usually say "I am experiencing." Everything else, however, is still present in awareness, though "dim" or "more peripheral." When I am attending to my wife, I still have a mind and thoughts. When I shift my position (a bodily activity) on the chair, it is present in the study. Awareness holds it all; awareness encompasses the whole of it.

The figure on the next page presents one view of this situation. Domains of awareness are demarcated regions established by boundaries and differences, and are specified by the nested squares. For most individuals in the West, the center of experience is the subject ("S"), located in the mind, which is located in the body, which is located in the world. The regions specified are similar in many ways to the partitioning of experience described in the TSK books (e.g., TSK, ch. 4).

Note that in the figure awareness is 'behind', yet pervades every aspect of 'what' is in awareness. To phrase this differently, whatever we experience we experience in, through, and as awareness. Usually, we experience it as 'awareness of', suggesting a subject who is aware of an object. This is the process called attention. Note, however, that I can be aware without being 'aware of'. There are many facets of awareness on which I do not focus attention, yet they present themselves to me. In the preceding example, my wife and her typing, my body, and the room were 'there' yet not in the focus of attention. The subject-aware-of-the-object relationship arises for those objects I attend to, but not for those simply in awareness. I can generate

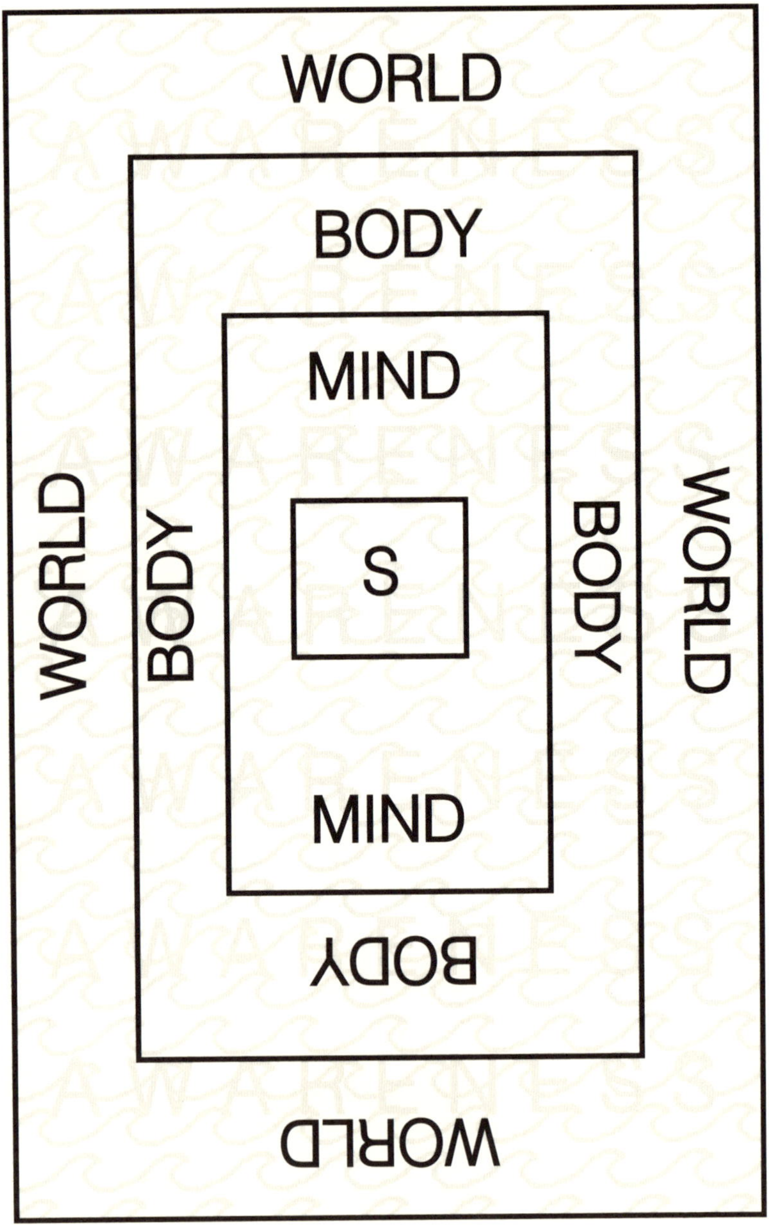
WORLD
BODY
MIND
S
MIND
BODY
WORLD
WORLD
BODY
BODY
WORLD

the subject-object relationship by focusing on these less focal facets of awareness and bring them to attention; I do not have to do this to be aware of them. They are simply present. As soon as I attend to them (or, using the previous language, focus my awareness on them), the subject-object polarity arises.

The subject appropriates as its own the awarenesses that emerge, but paradoxically, a careful tracking of what occurs shows that experientially it is the other way round. The subject is simply another kind of awareness. For instance, a thought emerges in mind. The subject notices the thought and appropriates it as "my thought." Furthermore, the thought (and its emergence) is called "my thinking." Yet the subject, who claims this mind as its own, arises in mind the same way thoughts do. Why should the subject be more primary than thoughts? The subject makes the claim to be more fundamental than mind, the "apparent source" from which thought and subject seem to emerge. But if awareness is partitioned into world, body, mind, and subject, why should one partition have primacy over another?

Most primary is awareness itself. Awareness conducts my mind, my thoughts, my body, and the world and its objects. All the various aspects of experience seem to be one kind of awareness or another; awareness links them all. Objects or things in the world, my body and all its sensations, mind and all its "contents" including the subject, all present in, through, and as awareness, each experienced as bounded, solid, or impenetrable. A critic might object that things exist outside awareness, but as will become apparent in the next section, objects arise only as experienced. The phrase "things are experienced" implicitly gives to those 'things' an existence, but the reality of that existence, from a phenomenological point of view, is one of the assumptions implicit in the "natural attitude," an assumption we have chosen to bracket or suspend.

Subject-Aware-of-the-Object

Although they are transparent to awareness, 'world', 'body', 'mind', and 'subject' are represented in the figure on p. 240 as opaque. This is because the substance of experience is usually considered impermeably bounded and separated into distinct objects. The world is solid; the body is solid: I cannot move my body through a wall. Most people experience a clear and unbridgeable separating of mind, body, and world.

One of the most pervasive of these distinctions arises when the subject attends to an object. The object could be a thought in the mind, a sensation in the body, or a sound, sight, taste, smell or touch that comes 'from' the world. As soon as the object arises in experience, so also does the subject. The subject and the object acquire 'substance' or 'solidity' when they appear in this relationship (see *Knowledge of Freedom* [T. Tulku, 1984] and TSK, Ex. 9, pp. 36–38).

This relationship founds the usual subject-object polarization and the 'coming-out tendency' of the subject (TSK, Ex. 9, pp. 36-38). It can be depicted in simplified form as follows:

S ⟶ O

The subject, discrete and separated from the object, attends to a particular object. As should be apparent from the "typing at the computer" example discussed above, the objects of my experience do not become objects-for-me until I actually attend to them. But what occurs when I am not attending? For example, most people have had the experience of concentrating so closely that they am not aware of the sounds around them, even as a diffuse non-focal background. During sports contests, an injured athlete will often not be aware of the pain until afterwards. The same can hold true for accident victims, who go into shock.

What seems most closely associated with not attending, not being aware, is focusing awareness tightly on a particular object of attention. There is something powerfully positive about this ability to concentrate the mind, which cuts like a knife. But the same capacity also reveals the tendency of consciousness to structure itself in restrictive and exclusive ways.

In the context of intention and will, we discover here the mental equivalent of overt action. Almost always, to focus awareness so narrowly and totally requires unwavering attention and, to persist, often requires will (though not in the case of the accident victim or injured athlete, where a natural focusing of attention is seamlessly integrated into the situation and effortless). At the other extreme, the conscious effort to concentrate almost always seems to disrupt concentration.

S ⟶ O

When the subject narrows its focus, subject and object are isolated and seem to exist as two discrete entities. Placing a box around the subject-arrow-object complex suggest this, implying that the subject-object seem to exist in a constricted world, excluding other 'things'. This establishes an inside, which is 'with' me, and an outside, away from and not me. There is, however, a more pervasive and subtle structure to experience that seems to underlie subject-object polarization. This is the topic of the next section.

The Ocean Metaphor: Linking Subject and Object

As a metaphor for the relation between subject and object, Tarthang Tulku (Moon and Randall, 1980) speaks of the ocean and its waves:

> Imagine that you live within the depths of an ocean; you are completely permeated by it. It gives to you, and you take what it offers, acting in ways that are expressive of the purity and power of the water. The results of your actions remain within that same sphere, flowing freely back into the water. But the 'ocean' is vast, unbridled power, not limited or constrained by anything, and constrains nothing. It permits everything, even ways of relating to it that are very limited and 'stand-offish'. . . . Perhaps you come to live

> on the very peaks of these waves and look across to the peaks of other waves around you. You pretend that reality is comprised only of what floats there on the peaks, that there is no 'underneath', not even any supporting water . . . . (pp. xxxi–xxxii).

These suggestions lead to modification of the relation between subject and object described above, as shown in the two figures below. At the top, the subject continues to attend to an object. Subject and object are separate and bounded, with 'attention' moving from the subject to the object. Neither participates in awareness. At the bottom, awareness is beginning to open. Now the experiential distinctness takes place against the background of a more inclusive awareness. The subject-arrow-object relationship 'sinks' into the 'ocean' of awareness. Subject and object are still distant, but now they are connected through and possibly as awareness. Even the arrow of attention is included in awareness.

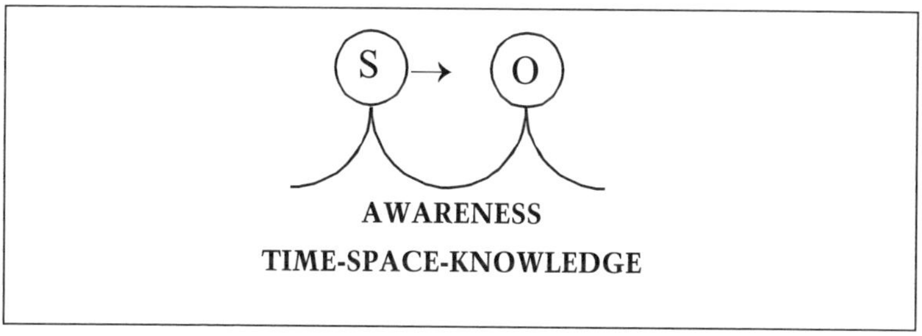

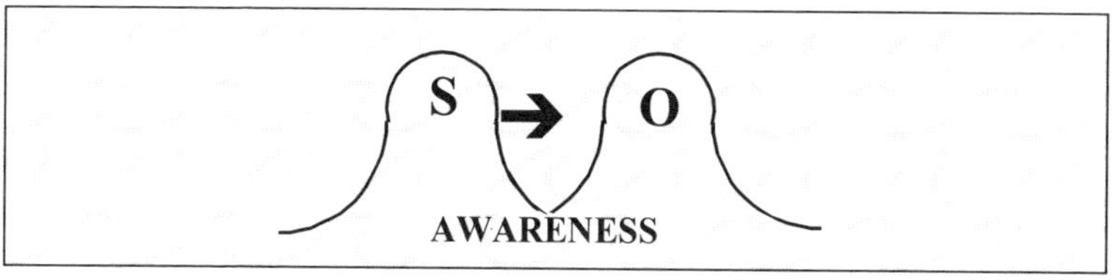

Opening Awareness: How Do We Notice When We Are Not Attending?

I follow the breath. My attention shifts away. At some point, I notice that I am not following the breath and I shift my attention back. This process takes place over time, as follows:

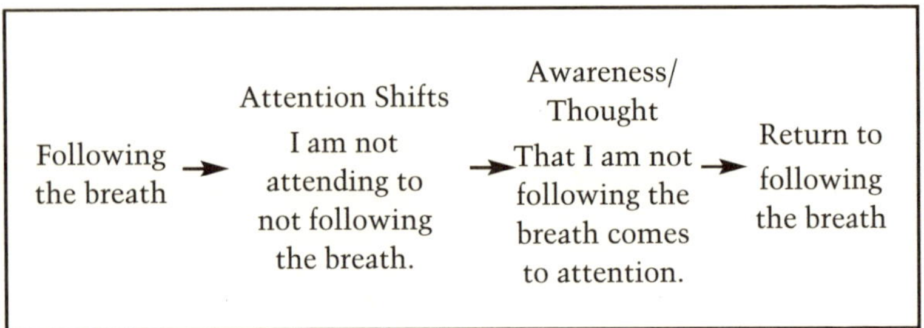

Awareness and Attention while Following the Breath

The critical moment is the transition from the unaware state, unaware that attention to the task is lost and non-reflective. Somehow there comes a new awareness—the thought, "I am not following my breath," which lets me continue the meditative task of following the breath. To do this, I must recall my original intention to follow the breath, and I must become aware of the discrepancy between my intention and what I am actually doing. Only then can I decide to bring attention back.

This recollection of the original intention arises on its own. It is not as if I do it. Something more than 'me' as subject, more than my self, seems involved. Awareness seems to respond on its own, independent of the subject's conscious efforts. One must, however, notice the recall when it emerges; otherwise, to state the obvious, it can have no effect. When awareness sends me the gift of knowing what I intended to do, the thought that I am not following the breath arises, and I can then acknowledge that I have ceased following the breath and bring awareness back to doing so.

This makes apparent another facet of continuing to attend. To achieve a goal or bring about an intention, I must remember that intention or goal, and I must continue to remember along the whole course of my progress. In the absence of that memory, progress toward the intended goal cannot be made. For instance, when I set out to cross the stream, I might become distracted by beautiful flowers, a rocky ravine or interesting wild animals and forget where I am going. I might even, on remembering, decide I am no longer interested; I might decide my current location is fine as it is and decide not to cross to the other side. Similar happenings occur when following the breath. I am distracted by thoughts and become immersed in them, forgetting my task. I might decide that what I am now experiencing is just fine as it is and decide not to continue following the breath. More likely, however, it will be the emergence of discomfort that grabs attention.

Bringing Something to Mind—Remembering

If I think "5 + 47= ?," I am aware of the intention to solve the problem, followed by the solution ("52!") emerging in consciousness. To clarify what happens in this simple process of bringing something to mind, here is an experiment. Please recall your telephone number. Now, attending closely to the process, recall your address. Next recall your date of birth. Repeat this sequence a few times, noticing how the recollection emerges in awareness. Note also how you conduct the particular recollection you intend (phone, address, date) and not "the wrong one."

If your experience is like mine, you will discover that the act of recalling is not something the subject actively does. I pose the mental question, "What is my phone number?" I shift my intention to make the words into a question about 'my' phone number. Suddenly, the phone number comes to mind. I realize that I have several phone numbers: my home and my office. I shift my intention, the focus of my question, from home to office to home to office. After each shift, a different yet correct number comes to mind.

Certain elements in this process are clear. It seems that I invite the memory of my phone number, as if opening to the memory as a mental possibility. This requires an intention focused precisely on remembering, since when I consider the question "What is my phone number?" simply as words, no recollection arises. Furthermore, the same words can access different phone numbers, my home and my office, but in each case I need to 'intend' recalling the one or the other.

Similarly, conscious events can often appear on their own, after which I appropriate them as mine. For instance, as I type a letter to a friend, I 'see him' in my mind, without the conscious intention to do so. My thinking about my friend somehow elicits from mind his mental image. Here a different kind of opening or inviting seems to be at work.

How does something come to attention? How do I bring something to attention?

Some experiences can 'force their way into attention' without being invited: a loud noise, a flash of light or an object that suddenly and unexpectedly comes into view. On the other hand, I can also shift my attention from experience to experience, and I can shift the focus of attention within current awareness or outside current awareness. Each possibility operates differently:

1. *Shifting attention within current awareness.* I look through my dirty office window at the building across the street. I shift attention to the dirt on the window and the building goes out of focus. I shift back to attending to the building and the dirty window disappears. Without shifting my eyes or focus at all, I can notice (attend to) the entrance to the building or the "No Parking" sign at the curb of the street. Another example: Someone walks across the street. My attention is focused inward, and while the person enters my view, I do not attend to her. Suddenly attention shifts, and I now notice her presence in awareness.

2. *Popping into attention from outside current awareness.* This refers to the sudden grabbing of attention, such as sudden and intense sounds, sights, or physical sensations. In the extreme situation, the new object of attention becomes central. But the same 'popping into attention' can happen with a more settled awareness, a non-grasping, open and receptive state of mind.

Of particular significance are the varying ways in which new events enter awareness. The 'bang' comes out of nowhere. The woman appears walking along the street. A pain explodes into attention. The cloud slowly forms and then dissolves in the sky. Thoughts and 'outside events' both emerge out of emptiness—a non-accessible openness. Later, they are no longer present.

3. *Shifting attention to outside current awareness.* I continue to look through my office window at the building across the street and follow the person who entered my view. It comes to me, as I notice her gait and the coloring of her coat and hat that there is a familiarity to her. In response to the familiarity, I realize that I know her, but I cannot recall who she is. She opens her purse, takes out keys and then passes out of sight. I shift forward to see the car she will enter since it might remind me of who she is.

Let us carefully track these processes. Up to the moment the woman disappears, most of what I notice is a response to what I am currently experiencing. When I act to bring a new object to attention—her car—I am ostensibly acting in order to attend to what is outside current (sensory) awareness. On the other hand, I was already aware that she was going to her car and "her car" was present for me in awareness as a possibility before I shifted my attention to perceive it. In other words, I shifted my attention to something that was already in my awareness, in mind but not in 'sensory' awareness. It is clear, however, that I may not have known the color of her car or its make, or even whether she was headed for a car or a motorcycle.

Another facet of this situation is the 'I' or 'subject' who does the shifting of attention. As noted above, to will something, to continue crossing the stream or watching the breath, requires directing awareness in a particular way. The most critical moment in each of these situations is bringing attention back to the task and keeping it there. In this process, the role of the self or subject seems critical. But how does this simple act of directing awareness arise in experience? This seems to be the essential phenomenon relative to plan, intention, decision, choice, and action. Each of these entails focusing or directing awareness.

We saw before that when I forget to watch my breath and then remember again, the breath is already in awareness before I shift attention back to it. Before I focally shift my attention to it, awareness of the breath, though not focal, is still there. In this regard, the subject could be described as a contained or partitioned region of awareness. When the subject attends to an object, the field of awareness contracts to include narrowly the subject-attending-to-this-object (represented above by the subject-arrow-object figure surrounded by the black border). Other experiences that might be in awareness are excluded. To attend to a particular, therefore, is for the subject to take a point of view that is restrictive and exclusive.

However, the 'ocean of awareness' is ever-present—through, behind, and as the subject-arrow-object polarity. Awareness, then, is broader than the subject. Awareness can track the actions of the subject, even its very emergence into awareness. Nor need the subject be actively involved (focusing attention) for awareness of a particular to arise. Awareness can be and is aware of particulars even when the subject does not focus attention on these particulars. This difference is like focusing on whatever is lit by the spotlights during the performance of a play, as compared to broader awareness of theater, audience, and stage when the house lights come up.

Action and Dreams

How does the subject narrow awareness? One way to begin to explore this question to consider physical action: the behavioral manifestation of focusing and choice. Suppose I make a noise, say by chanting "AH," and continue droning the sound for as long as I exhale. In this case, my intention to make this sound is directly perceivable in the world as a consequence of my action. The sound is from me, around me and within me, but it is perceivable by others as well. There is a subtle confusion or blending (depending on one's view) between me, what I produce, and what is not-me. The sound has been released to go its own way, both mine and not-mine at the same time.

Consider another situation: I intend and then choose to raise my arm in front of me until it is parallel to the ground. I see and feel it move. I recall choosing to move it. I have evidence of my choice and intention in the sensations associated with the arm's movement (kinesthetic, tactile, and visual). But now suppose I gradually slow down the movement so that there eventually comes a moment when it is so slow that no clear distinction can be made between sensation, vision, intention and movement/action. The clear demarcations that 'tell me' that 'I' am moving my arm are now missing. Intention, choice, and action (kinesthesia, vision, and sensation) blend into one.

This latter experiment (which I have carried out in Kum Nye and TaiJi, and also in TSK ex. 23, "Going without Going") clarifies a basic element of the intention-choice-action sequence. Time's unfolding is a prerequisite. The elements are separated in, over, and by time. In TSK terminology, they are 'timed out', a readout of time. In relation to achieving a goal (such as crossing the stream), my goal is not present but in the future, not 'here' but 'there'. The partitioning of experience into subject and object is matched by a partitioning into 'here-there' and 'earlier-now-later'. Slowing experience down alters this underlying experiential structure profoundly, which is why it can open up other possibilities.

Applying this insight to choice, decision, and intention is more difficult, since the subject-object polarization is ostensibly prerequisite for establishing an intention. Yet we can still say, at least in theory, that from the point of view of awareness itself, 'subject-object', 'choice', 'attention', and 'will' are all later elaborations by awareness focused on awareness. Using TSK terminology, they are a particular focus of Knowledge on Time and Space.

Dreams demonstrate in the most vivid way how awareness can constitute subject and object, complete with a credible world of objects having a history and a future. They reveal awareness as pervasive, able to constitute not only subject and object but also the whole structure of experience within which the particulars, including the subject, arise. Most telling and powerful is the persuasiveness and comprehensiveness of the dream experience. Within the dream, everything is presented in ways which the dreamer experiences as real at the time. In wholeness, complexity, and many other dimensions, there is a striking similarity between dreams and everyday experience.

Conclusions Concerning Awareness and Attention

A number of paradoxical or at least counter-intuitive, conclusions can be drawn from this discussion.

☆ Awareness does not require a subject who is aware. It arises independent of the subject.

☆ Awareness can be aware of something while also being self-aware. Co-present with the very process of being aware, awareness can be aware of how it is aware.

☆ The subject arises within and as awareness.

☆ The subject is associated with attention directed away from the subject.

☆ The subject appropriates awareness as its own. By claiming its primacy, the subject does not understand that awareness delivers to the subject its gifts without restriction or constraint.

☆ Attention often brings to light what is already in awareness.

☆ Awareness seems to open or to allow what emerges there. Attention and intention seem to guide that opening.

☆ New experience, inside or outside the body, seems to emerge from nowhere, from openness.

☆ Awareness can notice an infinite number of different qualities.

☆ Awareness seems carried along by the power of certain experiences.

☆ Whether the subject is consciously attentive or not, awareness continues to be aware.

☆ Awareness and subject can be co-present and co-aware.

☆ The 'subject' does not remember; rather, memory comes to the subject.

☆ The 'subject' opens the possibility of memory and mind; awareness might or might not respond.

☆ The subject's intentions conduct mind and allow the intended object to appear in awareness.

☆ Awareness can 'track' the very emergence of structures in experience.

Unaddressed by these conclusions is why consciousness does things one way and not another; why I do things this way and someone else that way. Are there other ways to be conscious, to 'do' consciousness?

For instance, can I attend without the subject who attends? Is it possible to attend in a balanced way that includes all facets of what is available to awareness?

Awareness, Attention, and Choice in TSK

The foregoing clarifies how the TSK exercises could have a profound impact on the way an individual experiences. Taking as given the phenomenological conclusion that all experience is constituted in consciousness or structured in a certain meaningful way, the exercises open experiential structures to greater variety and depth. Structures require boundaries, and the exercises open boundaries, so that they can both exist and not exist at the same time.

We also see why the TSK books do not emphasize the individual and the role of choice and will. These aspects of experience bring to the fore one of the most basic and constraining experiential structures, structures that the TSK exercises help us transcend. To emphasize will, effort, attention, or even awareness gives them an importance that would distract from the essential direction of the TSK books. The TSK exercises allow the subject to look past itself, to engage the subject while disengaging the subject, to discover that there is more than the subject at work.

As anticipated in the introduction, approaching experience phenomenologically and from an everyday point of view should nonetheless lead to the more inclusive vision expressed by what TSK labels "level-three" experience. Despite the particulars that comprise everyday experience, awareness (in phenomenological language, 'consciousness') pervades all experience. Discrete experiences are separated by boundaries, yet all are one with awareness or consciousness. We discover that experience simply arises, having no apparent origin. On the other hand, awareness or consciousness seems to respond to the intentions of the subject as if one can conduct the flow of experience.

We discover, as well, a certain quality of awareness associated with attention or intention. When attention is too tight or intention too effortful, experience rigidifies, freezing the inherently fluid and open nature of awareness. On the other hand, if awareness is not focused at all, the individual cannot function on a day-to-day basis. The middle ground allows but does not restrict. It might be described as effortless effort, open attention, or contentless awareness. It allows particulars to be as they are in the context of the whole and allows for distinctions while seeing unity. In short, it allows for a balanced way of being.

This open quality of awareness and intention is like holding sand. If I pick up sand with my fingers relaxed and loose, the sand slips through my fingers, and I am left with whatever grains are stuck to my skin. If, by contrast, I tightly grasp the sand, most of it slips out of my grasp. The solution is holding the sand in an open way, neither too loose nor too tight.

A well-known image for meditation, whose source I do not recall, makes a similar point. Imagine tossing a coin on a rubber sheet. The coin symbolizes whatever arises in experience, and the rubber sheet is the mind.

If the sheet is too loose (as when the mind is tired or groggy), the coin slides down into a well. If it is too tight, the coin bounces all over. But if mind is simply alert, the coin settles onto the sheet without bouncing or sliding. Awareness can concentrate narrowly and be very penetrating, or it can gently and evenhandedly hold everything. It can structure, contain, restrict, and exclude and at the same time open, change, and include.

A final question: Why do TSK exercises typically call for many repetitions? First, as a fact of experience, when we devote time and energy to an activity it has a momentum that carries over. Practicing the exercises repeatedly creates a momentum that can carry over into everyday experience. Second, we have practiced the everyday structures that have such power in establishing what occurs in usual experience again and again over our lives, beginning at a very young age. If one is to open these structures to alternatives, the new way of attending must be practiced frequently as well. Such repetition can create genuine opportunity and freedom.

Freedom is available. All options are open to us; our choices are unlimited. We must, however, consider the context for this observation. Freedom is available, change is possible, from one moment to the next. To know one's choices, to be open to this freedom, requires loosening the constraints that structure every aspect of experience. When we do this, openness and fluidity seem to arise naturally. When we do it most effectively and effortlessly, boundless love and joy arise from nowhere and attach to no thing. Action is spontaneous, and our lives are in harmony.

POSTSCRIPT

Considering what I have written, I sit, open to whatever comes to mind. The following emerges spontaneously in awareness; and I type the words as they appear.

Living beings are physical bodies of remarkable variety, with differing forms of existence. Every one joins with a partner to procreate, finds and then consumes nourishment, and uses pain as a cue to avoid threat and stay safe or to heal. Pain and illness come to everyone. All eventually die.

Within the vast arena of experience, there arises the subject dependent on will. As soon as there is will, there is a subject separated from object, separated from its intent or goal. At its most basic, I might have strong physical needs, say hunger or pain. With hunger, I move toward its satisfaction. I want it. In fact, were I not to eat I would die. With pain, I want to avoid the discomfort, though pain can indicate illness and be a guide for healing. In both cases, especially in situations of high intensity, the subject separates from the experience and wants to change it. I want that; I do not want this. I want to move toward that; I want to move away from this.

How then can the action of the self lead to the return of the self to oneness, to its very dissolution? Both the coming out of the self and never coming out of the self arise. Even though a self comes out, it has never come out. Even though it is always in one, it is always present as separate. The form the light assumes does nothing to change the light itself.

REFERENCES

Beere, D.B., 1980. "Phenomenology and Time, Space and Knowledge," in R.H. Moon and S. Randall, eds., *Dimensions of Thought: Current Explorations in Time, Space and Knowledge,* Vol. II. Berkeley: Dharma Publishing.

Beere, D.B, 1997. "Nonordinary Knowledge," in J. Petranker, ed. *Light of Knowledge.* Berkeley: Dharma Publishing.

Tulku, T., 1980. "Foreword: Ocean of Knowledge," in R.H. Moon and S. Randall, eds., *Dimensions of Thought: Current Explorations in Time, Space and Knowledge,* Vol. I. Berkeley: Dharma Publishing.

Tulku, T., 1984. *Knowledge of Freedom: Time to Change.* Berkeley: Dharma Publishing.

The Fullness of Space

Jack Petranker

We think of space as being empty, but in our ordinary interactions space is always full. To take the most obvious example, physical space is full of the physical; there is no room for anything else. (This is why science has such a hard time making room in the physical universe for consciousness.) Space is always full of content, and this is what makes it space.

Another way to put this is that space is always given to us in the form of a world. Each moment of experience arises within its own world, and space is full of this world. Space is full up. There is no room at the in.

This seems counterintuitive for the physical realm, because the content that fills it up is made up of two parts—everything that is and nothing at all—and we have a hard time identifying "nothing at all" as content.

So let's take an easier example: the world of a memory that comes to mind. If I call to mind a house I used to live in and some sense of what happened while I was living there, my memory world at once fills up with various items, each related to all the others. There are certainly perceptual memories, but there may also be emotional reactions, remembered feelings, events, and so on. There is a certain coherence to these items; they are linked in the same way that the notes of a melody are linked. Whatever they are, and however they metamorphose as I continue to reflect on them, they fill up the space of the memory completely.

If space is full, how can it allow anything else to appear? The answer is that it cannot. We think of something new coming along within a given space, but a more accurate view is that something new arises only when we shift spaces. We are not actually the ones who do this, though; instead, we shift along with everything else.

Space is always shifting. We often don't notice. For instance, looking around right now at the room in which I sit, space seems stable. But given what we know of physics, we would have to agree that in fact the room, and I myself, and the relationship between us, are all constantly in a state of flux. The fact that we usually ignore this, or perhaps cannot notice it, is like saying we cannot notice that the earth is hurtling through physical space at thousands of miles an hour. And the fact that when we do notice change, we think of it as change within a fixed space rather than space-shifting, is something like thinking of the sun moving through the sky, rather than the earth orbiting the sun.

There is nothing wrong with looking at things in this way, but recognizing that it is just one possible way of looking reminds us that we can train ourselves to see in another way. If we decide to do so, we will of course have shifted to another space.

Here again, looking at examples of non-physical spaces helps make the point. In shifting spaces, we not only shift from one moment of sense-perception to the next, but from a memory to a thought to an image to a perception to an emotional reaction, etc. These shifts are undeniable, even if we don't usually pay them much heed. That kind of space shifting goes on all the time.

If space is full, if it can hold nothing new, what does this say about our freedom to act or be differently? In one sense, we are completely boxed in; we have no freedom at all. On the other hand, there is still the movement involved in shifting from one space to another. Normally this movement goes on without our conscious control. (If you think you have conscious control over space-shifting, just try to keep the contents of your experience absolutely still for a minute.) Could we develop conscious control? Could this be the source of freedom? Perhaps, but at the least it would be an incredibly difficult thing to do.

A better approach seems to be to look more carefully at the process of space shifting. What allows the shifting? Does space-shifting itself happen 'within' another kind of space? In some sense, the answer must be yes: in order for anything to happen at all, there must be a space that accommodates it. This 'deeper' kind of space does not seem to be full in the same way that the space we experience is full, because it seems capable of accommodating an infinite number of spaces. In this sense, the space that allows space itself to shift never fills up.

How could we arrive at this space, or begin to explore it? Perhaps we are already there, since we do seem to have novel experiences all the time. Yet it all seems to happen without our being able to connect to it in a meaningful, knowingful way. This is a source of great frustration.

If we have had our fill of the full, the two-layer space-structure we have postulated suggests a way out. Whatever space we are in, we can question the contents of that space. We can question the context for each specific we encounter and the coherence of every pre-occupied space. We can make the presupposed problematic.

Of course, as we question, the content of the questioned changes, but that need not detain or concern us. We are not interested in tracking content accurately, but in discovering the context for all content. We do not wish to weave our way through space differently, but to unweave space entirely. Can we do this without fooling ourselves? Without stopping prematurely? Are we doing it any differently right now? How can we know? How can we notice?

Paradoxes of Transcendence in Time, Space, and Knowledge

Alan Malachowski

We should be very grateful to have
a limited body. . . . like mine,
or like yours. If you had a limitless life
it would be a real problem for you.
— Shunryu Suzuki roshi

Tarthang Tulku was once asked whether there is "some means of 'escaping' or transcending the conditions we find ourselves in." He responded (1998, p. 211): "[T]he idea of 'transcending' limitations or conditions is actually its own trap."

At first blush, this retort flies in the face of his own TSK writings. For these are replete with evocative descriptions of modes of 'being-in-the-world' where familiar restrictions on the correlative roles of time, space and knowledge allegedly have no place—where they seem, in short, to have been transcended.

There is no direct conflict here if we interpret Tarthang Tulku's suggestion literally. For he is telling us that it is the idea of transcendence (or the idea of that idea (1998, p. 211), rather than transcendence itself, that constitutes "its own trap." But this strict interpretation invites a further question. If the TSK project involves transcendence, can it avoid the paradoxes that transcendence itself apparently involves?

It is the latter 'hypothetical' question that I wish to explore in this article. Apart from the scene-setting remarks that follow, I will leave aside detailed consideration as to whether the TSK project really does have transcendent inclinations.

When viewed from any conventional angle, the TSK writings are extremely elusive—difficult to pin down and difficult to define. One reason for this, as Maxim Osinovsky confirms (1997, p. 376), is that Tarthang Tulku invariably refrains from providing any substantial positive characterization of the underlying TSK 'vision' (on the significance of this term, see Randall, 1997). He clearly prefers to say what it is not (e.g. it "is not philosophy, nor psychology, and not a religion," DTS p. xvi, and compare the long list of more detailed disclaimers that follows, pp. xvi-xx).

Systematic elusiveness usually rings the alarm bells for philosophers, and rightly so. But, in groundbreaking texts it is not necessarily a defect. The later works of both Heidegger and Wittgenstein are quite 'slippery' in this regard, yet they presumably gain at least some of their great potency from the fact that they are so resistant to orthodox interpretation. In this case, however, the degree of impenetrability involved does make the task of attributing to TSK any of the characteristics associated with 'transcendence' a risky venture—one which invites accusations of having misconstrued or misunderstood. Be that as it may, there are at least two good reasons for making an explicit 'TSK-transcendence' connection.

The Transcendence Connection

The first reason is textual. The TSK book, Tarthang Tulku's original vehicle (cf. Randall, op.cit.) for the written expression of the TSK vision, offered a complex recipe for self-transformation in which normal forms of mental and physical 'constriction' are left behind (so we might say: 'transcended'). Moreover, this 'canonical' text employed an explicit vocabulary of 'transcendence': "The real issue is to transcend the observed isolation of the self. . . . We need to transcend the 'isolated self perspective altogether." (TSK, p 66)

The second reason for associating TSK with transcendence flows from the legitimate expectations it is liable to generate in that connection. This applies to the TSK literature as a whole, but particularly to the rhetoric deployed in attracting the attention of new readers. The back cover of the TSK book is typical in the latter respect. It promises (in some printings, at least) "a significant breakthrough in liberating our perspectives on knowledge and human possibilities." When faced with such radical claims, readers are entitled to expect something extraordinary, something that engages their highest aspirations—such as the desire to transcend the confining aspects of their existence.

Not only that, but the TSK project on its side needs to engage just such aspirations if it is to differentiate itself from other ambitious projects in the field of self-transformation—a field where, as Osinovsky (1997, p. 379) observes, "there is no shortage of recipes." Why should readers take the trouble to tackle the considerable intellectual and practical demands of the TSK material, if despite surface rhetoric to the contrary, it does not even attempt to help them transcend what they perceive to be constricting aspects of their relationship to time, space and knowledge?

If we adopt the provisional strategy of interpreting TSK as a transcendence-seeking project on these grounds, the most pressing question becomes: "Does TSK actually deliver the goods; namely, transcendence?"

This question is not at all easy to answer. The practical results of working with the TSK written material and its designated 'exercises' are difficult to quantify or otherwise assess, independent of the testimony of the participants concerned (Cf. Petranker, 1993, chapters 1, 7, and 8). Such problems of external verification are presumably inevitable, because the 'knowledge' that TSK work is intended to make available "is not a matter of contents, but of active knowing expressed in inquiry itself." (LOK, p. xvi). As such it cannot be treated as "an 'end product' that can be viewed in isolation from the contemplative processes which enable us to embody it." (Malachowski, 1997, p. 227)

By the same token, it is difficult to determine the degree of common-sense realism entailed by TSK-related claims. When the texts and personal accounts of TSK practices speak of "healing," "unlimited knowledge," "translucent boundaries," "immortality," and the like, do they mean to imply what they appear to imply: that, in virtue of their TSK-related efforts, TSK practitioners are likely to enjoy perfect health, know everything there is to know, gain the ability to walk through walls, live forever, and so on? If so, the reader is justified in asking for objective evidence that such unprecedented levels of attainment have been reached, but to my knowledge evidence of this sort has yet to be made public. If there is no such evidence, the reader is warranted in querying the very use of terms like "healing" in the first place. What meaning can such words retain when they are detached from their conventional connotations and their practical consequences?

The general concern here is that because the TSK project has been set up to be 'omni-compatible', its contentions cannot be refuted, even in principle. More specifically, this means TSK is vulnerable to the well-known Popperian objection that it has no empirical content, and hence no practical application. The guiding thought is that if there is no possible Y such that the truth of "Y is the case" implies the falsity of "X is the case," then the claim "X is the case" excludes nothing, and in ruling nothing out, has no determinate sense.

The TSK writings themselves appear to compound these dilemmas by equivocating between 'realism' and 'phenomenology'. When a 'realist reading' is encouraged, a typical TSK-based contention fleshes out something like this:

> When it it claimed that under TSK-scrutiny, boundaries between otherwise separate and solid objects appear to dissolve, they really dissolve—because that is how things really are in TSK-perceived reality.

By contrast, a 'phenomenological reading' of the same sort of contention conjures up first-person observations that stop short of realistic implications for the extra-experiential world:

> When it is claimed that under TSK scrutiny, boundaries between otherwise separate and solid objects appear to dissolve, this is a purely subjective consideration—simply a private matter of experiencing things as if they are such and such, and it does not in any sense imply that they really are such and such (e.g., "unbounded").

And, when, as in the following example, the text itself fails to signal which of these interpretations, if either, is appropriate, the reader is left in a frustrated state of suspended hermeneutic animation:

> This does not mean that 'we' can walk through walls. But knowingness can apprehend an uncharacteristic sequence of time, which would ordinarily be considered as a 'passing through the wall' sequence. (TSK, pp. 197–98)

The Priority of Transcendence Over Verification

I suspect that these thorny philosophical problems of 'verification' are considered to be merely symptomatic of 'lower level thinking' by the TSK fraternity. Perhaps this explains why, apart from the few exceptions flagged in earlier references (Osinovsky, 1997, and Petranker, 1993, chapters 1, 7, and 8), such problems have not been taken seriously in the TSK literature. But unless such problems are satisfactorily resolved, members of the wider community are likely to mirror that complacency by refusing to take TSK itself seriously. The epistemologically sophisticated culture of the West (Malachowski 1997, *passim*) furnishes them with plausible motivation for sticking to a TSK-resistant line, more or less as follows:

> The kind of thinking that requires that putative assertions about the nature of things must satisfy definite, community-wide criteria of legitimation is the only coherent and worthwhile form of thinking we are presently capable of.

From this standpoint, the onus is on the advocates of TSK to show why such a view can be dispensed with—why well-proven epistemological standards ought to be liberalized on TSK's behalf.

Important though these issues may be, our initial 'hypothetical' question is more interesting. For it provokes the thought that TSK could be susceptible to criticism even if the verification issues are resolved in its favor. That is to say, TSK might be deeply problematic even if it could be demonstrably seen to deliver the appropriately transcendent goods. Or to push the point a bit further: its very success on the transcendent front might turn out to be the Achilles' heal of TSK. That is the issue on which I now focus, although the discussion that follows also has some bearing on the issue just raised: the problem of whether the verification issues surrounding TSK can be resolved.

Paradoxes of Transcendence

What my argument urges us to reject as incoherent is the aspiration to leave behind altogether the constitutive conditions of our humanity, and to seek for a life that is really the life of another sort of being—as if it were a higher and better life for us.
— Martha Nussbaum (1990, p. 379)

The argument that Nussbaum is making here falls in with a presumption in favour of naturalism that now pervades intellectual endeavors in the West. A physicalist ontology provides the starting point: "[O]ur world is a physical world and all the things that really exist are 'nothing but' physical" (Stich 1996, p. 195). From there, whatever appears to depart from the realm of the physical is brought back under its umbrella by tying it to something physically observable. Here, the value and significance of all aspects of human existence is to be explained in behavior-relative terms—as something that can only be assessed within particular customs and practices of particular human societies, materially circumscribed by forces of nature. There is no higher, culture-free, ahistorical, nature-independent court of appeal. Hence the good life for human beings is essentially a life lived in accordance with cultural guidelines, organically evolved—or, on occasion, methodically constructed—in the relevant societies and physical surroundings.

This naturalistic picture tries to demystify our human existence. As Dews (1995, p. ix) points out, it "frequently presents itself as an open-eyed acceptance of the cold reality of the modem world." It also has to allow for (because it cannot consistently rule out) a plurality of possibly conflicting conceptions of how societies should organize themselves and how individuals should live their lives.

In the first stages of the argument cited above, as reconstructed for our purposes, Nussbaum identifies a salient factor common to various forms of social organization. Although the content of the naturalistically-sanctioned, humanly good life may vary enormously according to circumstances, such a life will always be lived within the confines of some constraining features of culture and nature (following Nussbaum, we will concentrate on natural constraints). A life lived beyond any constraints of this kind could never be a truly human life. In its second stage, Nussbaum's argument then tries to establish that the humanly good life so defined is the best possible life for us, if not, indeed, the best life as such.

Nussbaum's argument is presented in a complex and rhetorically diffuse form. But, when the disparate elements are brought together and unpacked, they provide some interesting points of contrast with TSK. Nussbaum wants to show that:

(1) The humanly good life is necessarily subject to constraints;

(2) Although it is often sensible for us to try to overcome those constraints that are "constitutive of our humanity," complete success in this respect would be self-defeating;

(3) The idea that a human being might lead an unconstrained life, beyond all limits, is incoherent;

(4) Beings that could lead an unconstrained life might well be worse off than human beings precisely because their life lacked limits (compare the quote from Suzuki at the head of this paper).

Some brief comments on each of these four points will help prepare the ground for an examination of the contrasts with TSK.

The Necessity of Constraints

The riddle that the human being is supposed to solve can be solved only in being what he or she is, and not in being something else, in the immutable. — Nietzsche

What lifts (1) above the level of platitude is the penetrating light it sheds on common fantasies about the best possible kind of life for human beings. When faced with the brute, life-constraining facts of our physical vulnerability and mortal finitude, we are frequently inclined to wish away those facts in a fantasy of omniscient 'ultra-freedom'. If only our friends and loved ones could be immune to illness and the ravages of time and chance; if only they could live forever! But, we do not stop to consider how much of the deeper meaning and significance of our lives would vanish if our wishes for a better life in this 'less-constrained' sense were to come true. Consideration (1) encourages us to confront the issues at stake here. Consideration (2) reminds us of their paradoxicality.

Overcoming the Constitutive Conditions of Our Humanity

It is a good thing for us to strive to overcome the harsh limitations of our particular brand of physicality. The potential rewards are obvious: a better and easier life. Paradoxically, however, success in this sphere could be more destructive than the original limitations. To see why, think of mundane examples: the ambitious weightlifter who somehow overcomes physical limitations on the human frame so that she can lift weights of any magnitude, or the keen high-jumper who acquires an ability to defy gravity. In such cases, the activities concerned would no longer make sense, and the motivation to pursue them would drain away. As Nussbaum (1990, p. 371) herself puts it, there could be "no athletic excellence at all, and no meaningful concept of athletic excellence, in a life of a being that is, by nature, capable of anything, and physically unlimited."

But Consideration (2) does not apply to just any old set of limitations. It picks out those which are supposed to be "constitutive of our humanity." And as such, it provokes us into considering which kinds of constraint, if any, are an essential part of our human identity—part of what Nussbaum (1990, p. 366) calls "the whole human package." This kind of reflection may help to reveal the self-defeating nature of transcendent ambitions.

The Incoherence of Ultra-freedom for Humans

Coercion from outside, strong temperamental inclinations and passions within ourselves, do nothing to affect the essence of our freedom. They simply define its action by imposing limits. They give it a peculiar character of its own.
— Thomas Merton, 1956, p. 12

Consideration (3) delivers the final blow to the fantasy of ultra-freedom. A life without constraint would be an empty life. Since there would be no meaning/value-shaping reference points in this life, a being living it could not be recognizably human. Such a being could have no "peculiar character of its own," and its existence would have no value in human terms. Hence, as far as humans are concerned, the notion of 'ultra-freedom' is unintelligible. Lest this fantasy-exploding conclusion seem too pessimistic, Consideration (4) offers some consolation.

Why Transcendent Beings Might Get Left Out in the Cold

Because I know that time is always time
And place is always and only place
And what is actual is actual only for one time
And only for one place
I rejoice that things are as they are.
— T.S.Eliot, Ash Wednesday

We seem to be incapable of conceiving what the life of an unconstrained being would be like. For without building in some restrictions, we can get no fix on the nature of the possible activities of such a being, and no measure of its likely motivation for acting in the first place. The claim voiced in (4) consoles us for having to live a constrained life, by suggesting that even if an unconstrained life were possible, it might not match up to a humanly good life. It might fall short because none of the worthwhile ends that human beings can attain in virtue of their inherently restricted setting would be achievable. Viewed in human terms, the unconstrained life is always going to be a life bereft of value.

Qualifications linger here ("might not match up," and so on), in tacit recognition of a corresponding tentativeness in Nussbaum's approach. Perhaps she is merely being cautious because she wants to avoid the overconfident dogmatism and blinkered reductionism that the 'culturally internalist' outlook of naturalism can sometimes foster. But, it may also be that she has deep and far-sighted philosophical qualms about the ultimate jurisdiction of human-centered verdicts on the nature of the good life. If an unconstrained life is possible despite being inconceivable by us (and, *a fortiori*, for us)—if it is *metaphysically* possible—then such a life might appear to be better than the humanly good life from its own unconstrained perspective. Whether it would actually be better is a deliberately moot point in Nussbaum's discussion.

To say that the humanly good life is better on its own terms than a kind of life we canot conceive (and which is therefore most likely impossible anyway) may seem vague consolation for having to endure constraints. But we can take it as a contribution to a wider process of reflection that strengthens the reconciling hand of naturalism. For when we examine the part that certain constraints play in our lives, we find we can enhance the stoical stance that naturalism has hitherto tended to adopt. That is, we can do more than resignedly make friends with some of the limitations on the human condition. We can exploit their 'enabling' powers to the full, and we can celebrate their capacity to keep us out of a transcendent void of values. Then, along with Eliot, we can "rejoice that things are as they are."

Towards a Critique

By learning to directly contact the essence of our being we can discover an unbounded freedom that is not only freedom from some external constraint, but is itself the dynamic expression of the meaning and value of being human. (TSK, p. xxxv)

Imagine a perfect TSK practitioner—someone we could aspire to be like—who embodies the TSK teachings to the fullest extent and who has "directly contacted the essence of her being." What qualities would we expect this person to display, given the claims made in the TSK literature (leaving aside complications concerning whether such qualities could only be "seen through TSK-oriented eyes")? Here is the sort of answer that seems to me licensed by the TSK writings:

> A perfect TSK practitioner would be a purely spatial being—and self-consciously so. All the kinks in her existence would have been smoothed out, all the limitations and 'friction' dissolved. She would shimmer with a self-illuminated, all-embracing 'knowingness' that leaves nothing epistemologically opaque. Finally, she would have access to infinite resources of time and be acutely sensitive to its rhythms, while also appearing to live an unhurried, timeless existence.

Is such a characterization credible? And if it is, are the qualities it describes really the kind of qualities human beings would benefit from having?

Nussbaum's anti-transcendence argument raises doubts as to whether anything like this characterization of a perfect TSK practitioner is credible. At the same time, they make it highly disputable whether the qualities so described would even be beneficial to human beings.

A purely spatial, epistemologically omniscient, timeless being would be a radically unconstrained entity. It would enjoy "unbounded freedom." However, since on Nussbaum's understanding such a being could have no distinctive 'shape', and hence no value-encapsulating contours, it would lack key "constitutive conditions of humanity." Hence, our characterization of a perfect TSK practitioner can have no credibility as a characterization of a human being.

But is it without credibility *per se*? Is there no kind of being that a phrase like "purely spatial, epistemologically omniscient, and timeless" could meaningfully refer to?

Nussbaum's position does not yield an unequivocal answer to such questions, but it stacks the odds against credibility emerging intact. Perhaps a godlike being could fit the bill, but in the end we have to ask ourselves: "Just what is it we are imagining when we believe we are imagining an ideal 'time-space-knowledge' being?" An intelligible, non-circular answer is difficult to envisage.

Again supposing our characterization of a perfect TSK practitioner could somehow be made credible, would the qualities it describes be beneficial qualities—the sort of qualities human beings would be better off possessing? Again, Nussbaum's insights raise grave doubts. For the sake of brevity, let us focus on the question of 'spatiality'.

On a naturalistic reading, space is a paradigmatic value-neutral phenomenon. It is the things occupying space, the relationships between them, and their activities that merit attributions of value. In a purely spatial universe there would be nothing to generate or embody value—and nothing for value assessments to get a grip on. If, after an equivalent of the big bang, there had only been space, then our universe would have been a value-empty universe. It is the advent of space-inhabiting creatures who develop socially-coordinated value-responses to their environment that ushers in the possibility of there being anything worthwhile in the world. If so, then were it feasible for such creatures to transform themselves via TSK into purely spatial beings (reveal their underlying 'spatial nature'), doing so would again flatten the value of their existence down to nil.

Hence, it is much better for us to attempt to live a humanly good life than to try to dissolve our identity-constructing limitations in space. Pure spatiality can confer no genuine benefits upon human beings. Clearly, if these concerns prove valid, they make it doubtful whether the TSK project is at all viable.

Transcending the Possibility of Critique

In the end, the naturalistic suspicion we have described comes down to the following worry: To the degree that it is successful, our pursuit of transcendence ineluctably induces alienation, thereby making us less human. Are such worries justified when it comes to TSK? Can the critique be sustained?

There are a number of ways to defend the TSK project. First, it would be relatively easy to get TSK off the hook on a verbal technicality. TSK afficionados could readily cite textual evidence which seemed to contradict claims and distinctions made above. They could, for example, take issue with the alleged 'realism/phenomenology' dichotomy, perhaps on the grounds that TSK avowedly distances itself from any pretensions to tell us 'how things really are' (see, e.g. TSK, p. 124: "What is being suggested is not an absolute position describing the truth about a world order, but rather a way of growing without ever falling into a stagnant orientation.") But such a defense could be marshaled against just about any line of criticism of TSK, feeding doubts as to whether the TSK texts have any fixed meaning, and making us wonder whether the 'elusiveness' of the TSK vision we referred to earlier is more harmful than we intimated. Furthermore, this kind of verbal jousting goes against the whole spirit of the TSK enterprise. And it leaves the substantive issues untouched.

A more refined version of this move would be to develop a 'double ontology' conceptual apparatus: one level for conventional reality and another for TSK-founded reality. Buddhist philosophy has favored a similar move to overcome what we might call 'The Bodhisattva's Paradox':

> Bodhisattvas are beings compounded of the two contradictory forces of wisdom and compassion. In their wisdom, they see no persons; in their compassion, they are resolved to save them. (Adapted from Conze, 1959, pp. 129–30)

This paradox is resolved by distinguishing between 'conventional' and 'ultimate' truth (the well-known doctrine of 'two truths'; cf. Newland, 1995). While on the conventional level people exist and their suffering is real and should be alleviated whenever possible, ultimately there is no such thing as a person, and hence no such thing as personal suffering. Similarly, the TSK writings explicitly recognize that "the world of conventional experience depends on mutually limiting entities" (KTS, p. 472) and imply we can have access to a world beyond 'limitations'. Adopting a parallel position (i.e., TSK-truths are 'true-for-the-time-space-knowledge-dimensions-of-reality' but not 'true-for-conventional-reality'), would eliminate some of the ambiguities in the primary texts. But it would still provoke Nussbaum-like doubts as to whether the level of reality invoked is something human beings could and/or should aspire to.

Naturalism recommends that we give up trying to climb outside our skins and begin, instead, to make ourselves more at home in everything we are able to seamlessly create out of our natural endowments: our bodies, our history, our culture, and our physical environment. Here we have the seed of a more powerful response to the critiques advanced above. Experience is also part of our natural endowment, and the TSK texts suggest that when we explore the nature of our experience, we may find that its restricting features are self-imposed rather than simply 'given'. Although its results may threaten to overturn conventional notions of what experience is supposed to be, this kind of 'TSK exploration' is perfectly compatible with naturalism.

Can the TSK project can be rescued in this way—by a resolutely phenomenological or 'experiential' interpretation of its aims? While this would mean divesting TSK of any claims about the nature of extra-phenomenological reality, it could lead to some interesting modifications of the otherwise problematic methodology of traditional phenomenology (cf. Beere, 1980). For the larger community, however, this might strip TSK of its unique interest. For there are any number of opportunities on offer to facilitate 'transcendence' within the realm of subjective experience alone (think of drugs, contemplative techniques, virtual reality, and so on).

How can TSK show that it does not involve a withdrawal into mere 'subjectivity' in the face of the very real problems facing us in the external world? If we set aside verbal disputes, double ontologies and the possibility of a retreat into phenomenology, perhaps a more radical approach could demonstrate that TSK is able to operate beyond the reach of the sort of criticisms we have mentioned. Taking its cue from blind spots in the naturalistic outlook, TSK could define itself more instinctively as a radically new venture, one which is therefore not necessarily answerable to the demands of naturalism. In today's climate, this would amount to claiming philosophical autonomy. Let us conclude by briefly explaining what such a strategy might look like.

Naturalism attempts to explain four main things:

(1) What there is;

(2) What we can do with what there is and why this is possible;

(3) How we can know what there is;

(4) How what there is came to be what it is.

It tries to wrap these four kinds of explanation into a single, homogenous package that renders our world familiar on naturalism's own epistemological terms.

Scientists have been so good at convincing us they can provide such a package for everything that other intellectuals, including many philosophers, have jumped on the bandwagon. Any account of reality has to fit neatly into the package of naturalism, or, failing that, to provide its own equally encompassing package. So, for example, an account will be inadmissible if it refers to entities X, Y and Z without being able to answer the question "How can we know that there are such things as Xs, Ys and Zs?").

However, the naturalistic enterprise has so far been unable to cope with a number of fundamental questions—think, for example, of questions concerning the ultimate origins of our world ("How was it possible for our universe to come into being?") and the place of consciousness within it ("How can consciousness emerge from matter and space?"). The current scientific world view is often expounded in terms which make it seem almost complete ("We know more or less what the universe and its contents are like; we just need to fill in a few more details to get the whole story straight.") But such appearances can only be maintained by ignoring key issues about both cosmology and consciousness. The knowledge that science produces currently rests on a largely invisible 'cloud of unknowing'.

When these issues are highlighted, the present scientific picture looks far from complete. It could turn out that both the universe and its sentient inhabitants are at present drastically underdescribed. For instance, perhaps the TSK-recalcitrant view that space is epistemologically inert and value-neutral (assumed in the discussion of the spatial qualities of the ideal TSK practitioner above) is mistaken. In any event (so the strategy would maintain), it is probably premature to talk about "the constitutive conditions of humanity," because human potential could outstrip the boundaries of what has been behaviorally manifested as a norm in all the cultures that have existed so far, and in this way could be culture-transcendent.

Against this background, TSK's failure to fit easily into the naturalistic package does not by any means make it a doomed project. The deep problems with naturalism generated by its results so far suggest that naturalism could be inherently flawed, that TSK need not be answerable to it, and that TSK is entitled to attempt to 'go it alone'.

To succeed as an autonomous venture, the TSK vision still needs to meet some very general external requirements. It has to foster a sustainable 'way of life', one which is open to people from very different cultural backgrounds. And this 'way of life' has to embody its own 'verification practices'—otherwise anything could be sanctioned as a TSK option.

Such developments presuppose a shared vocabulary. The primary texts have presumably set a useful precedent here, and the linguistic tools thus provided could be further refined by individual TSK practitioners as they respond to their experiences of interacting more directly with time, space and knowledge. Perhaps one of the main reasons why the TSK project now looks so problematic from the outside is that this process of 'language building' is still in its early stages.

If a TSK 'way of life' emerges fully grown, the kind of concerns voiced earlier will either become irrelevant, (because TSK has established a new, self-standing outlook in which they play no part; as it were, human history simply moves into a fresh phase within a new 'living paradigm of transcendence') or TSK will have constructed its own way of dealing with them. Whatever the outcome, this article has tried to raise questions which will help clarify the direction that the TSK project takes over the next few years.

REFERENCES

Beere, D., 1980. "Phenomenology and Time Space, and Knowledge." In Moon, R.H. and Randall, S. eds., *Dimensions of Thought: Current Explorations in Time, Space, and Knowledge*. Berkeley: Dharma Publishing.

Conze, E., 1959. *Buddhism: Its Essence and Development*.New York: Harper and Brothers.

Dews, P. 1995. *The Limits of Disenchantment: Essays on Contemporary European Philosophy*. New York: Verso.

Malachowski, A., 1997. "Epistemology in Paradise: The Spatial Embodiment of Knowledge and Value," in Petranker, 1997.

Merton, T., 1956/1993. *Thoughts in Solitude*. Boston: Shambhala Publications.

Newland, G., 1995. *The Two Truths*. Ithaca, NY: Snow Lion Publications.

Nussbaum, M., 1990. "Transcending Humanity," in *Love's Knowledge: Essays on Philosophy and Literature*. New York: Oxford University Press.

Osinovsky, M., 1997. "'Hard Knowledge' and Time, Space, And Knowledge," in Petranker, 1997.

Petranker, J., ed., 1993. *Mastery of Mind*. Berkeley: Dharma Publishing.

Petranker, J., ed., 1997. *Light of Knowledge*. Berkeley: Dharma Publishing.

Stich, S., 1996. *Deconstructing the Mind*. New York: Oxford University Press.

Shunryu Suzuki roshi, 1970. *Zen Mind, Beginner's Mind*. New York and Tokyo: Weatherhill.

Tarthang Tulku, 1998. *Teachings from the Heart*. Berkeley: Dharma Publishing).

My thanks to Jack Petranker for his valuable editorial suggestions.

The Significance of 16

Tarthang Tulku

As far back as we can trace in recorded human history, and most likely for millions of years before that, we have been proceeding from the perspective of the 'I'. The 'I' is a position we take, and from that position we readily identify and take on an identity. Once identity is in place, we find ourselves 'here', a unique location that sets us in opposition to a world that we then set about knowing and acting on.

If we want to investigate the 'I' perspective and explore alternatives, a good starting point is to explore the structures that support it. Turning first to the 'here', we can say that 'here' is located in space. This means it stands in relation to other possible points in space, which we regard as being 'there'. In this relationship, a 'from/to' structure operates. The 'I' as the source of knowing and acting, is located at the 'from', moving out 'to' a world located elsewhere. The 'from' depends on taking a position, but also makes the act of 'taking a position' possible. Once I have identified a perspective as 'mine', I am ready to point out 'from' as the root or source of that perspective. 'From' is my home, the space I occupy. It lets me present something 'to' someone, or have something presented 'to' me.

'From' and 'to' lack all content and substance, and this makes it difficult to inquire into their operation. But even when content and substance have not yet been established, it is possible to investigate structure through the medium of symbols. Our focus on 'from/to' allows for a new possibility: seeing 'I' as just such a symbol. In this view, 'I' is a tool we use to take hold of 'from' and 'to', and thus to present the self. Every culture and each language seems to make use of this same symbol to accomplish the same purpose.

'I' is unitary, located 'here' and not 'there'. But 'I' is also bipolar, depending on both 'from' and 'to'. In this way, 'I' as symbol expands to encompass two points. It establishes the 'point-being' of those points, which in turn—in light of their interconnection—could be seen as symbols for the possibility of communication.

The symbol of the two points likewise points beyond communication toward existence. Point-being establishes 'my' subjective existence, together with 'my subjectively experienced world'. Whether I express the two-point structure through gestures, verbal communication, energy and action, or the operation of the senses, I am naming myself and my world and asserting our mutual existence.

In one sense, the dynamic of this establishing is wholly subjective, completely my own. At the same time, it depends on two or more persons agreeing to accept and share the same symbolic structures, not only now, but as a fundamental, stable order. To be a world, the world must be present directly for me and also for the other.

For example, acting as 'I' we may engage in some kind of creative act or experience something very personal, such as joy, sadness, or confusion. But even though our starting point seems intensely subjective, it has to move out or connect in some way. This is what it means to say that the 'from' must have a 'to'. If we speak, there must be space to conduct the sound of our voice; if we act, our action is directed toward some object. Even experiences that seem disconnected, utterly private, or completely spontaneous and unexpected still link up to a 'to'. And this is so even if their way of linking up is to deny that any link is in operation. Language, objects, tools, sensory perception, patterns of experience and behavior, the data given to the senses, the tone or mood of experience: All these begin in the same way and end in the same way. We start by establishing our position as the 'from', and we end up by affirming a relationship to a 'to'. To repeat, 'I' is the symbol for this unfolding, which affirms both self and world as real and existent 'points of being'.

'I' and Zero

We have seen how the symbol of 'I' goes from one point to two, and how the symbolic structure it evokes becomes the 'points' of being. Is there anything outside this structure, anything that does not start out from the symbol 'I'? In terms of the structure we have been investigating, the answer seems to be that nothing at all can escape this founding structure. Yet this 'nothing' does not have to be empty. We are operating at the symbolic level, and 'nothing' can also have its symbol. As the fitting symbol for the 'nothing' that escapes our 'being-point' structure, let us use zero.

The availability of zero as a symbol immediately allows for a different question. When the 'I' establishes a being-point, what is that point made of? At the level of symbol where we are operating, the right answer seems to be 'nothing at all'. But because zero is the symbol of that nothing, we are free to say that the established point is zero. 'From' is zero and 'to' is zero as well.

Now, can one zero-point be connected to another? It is hard to imagine how the lines of connection could be drawn. Yet when it comes to 'from' and 'to', we somehow manage. Even though we are moving from zero to zero, with no inside and no outside, we set the operation into motion and follow it out, until in the end the 'I' 'here' has linked up to 'something' 'there'. Only because of this link can we speak of a being-point 'being there'.

What do we actually establish when we move out from the zero-point of 'from' in the direction of 'to'? If the first point, the 'from', is zero, establishing that point means establishing the zero as symbol and representation. In this move, nothing at all has been established, yet all the structures we rely on as being-points—'from' and 'to', 'front' and 'back'—have been set in place. As the zero carries on, representing nothing at all, it manages to present the object as presenting itself. True, this presentation remains zero, but this lack of substance does not seem to undermine its capacity to set up and set in place.

Zero as symbol seems to point in two directions. Viewed 'externally', as nothing at all, it is the symbol for the quality of space. Viewed 'internally', as the nothing of the being-point, it is the symbol for all perceptions. Nor does this exhaust the power of the zero, for zero as symbol makes possible calculation, whether in the number system, the programmed logic of computers, or the measured-out grid of geometry. More fundamentally still, zero is linked to communication—to the move that links 'from' and 'to' in a meaningful way. For if 'from' and 'to' were not separated by a gap—zero in another guise—the need for communication would never arise.

Such speculations lead to an unusual thought: All that we are, including our mind, our thoughts, and our existence, can be represented by the symbol of zero. In this way of thinking, zero becomes the master symbol, the symbol of symbols. It encompasses every manifestation of every symbol that helps to establish the 'from/to' duality.

For instance, zero as symbol gives the possibility of naming, an action grounded in the power of symbols. From naming comes language, from language come qualities, and from quality come distinctions. With distinction come all the points of every realm, from emotions to perceptions to planets, galaxies, and the periodic table of the elements. In a rush of creativity, all existence is presented; all existence comes forth.

Zero and One

If zero is the symbol of both nothing and everything, 'one' is already present in the symbol 'I'. The 'I' is one in multiple senses: single, unique, and first to appear. It is the first case, the first instance of what is, the content and content-giver for the twofold being point.

'I' presents itself as central. Without zero, the best we can say about this may be that 'I' manifests in some mysterious way out of an unformed 'everything', and that it stands in relationship to all that comes after. But the availability of zero lets us see it somewhat differently. Within what is undifferentiated, and therefore zero, a zero point of view emerges, becoming 'I', the one alone.

From the one of 'I' and the operation of the zero point, the well-established order of 'from' and 'to' arises, becoming subject and object. 'From' and 'to' establish a baseline, and communication along that baseline manifests in the twofold reality 'my . . . world'. Depending on the operation of the six senses (the five physical senses and the mind), the distance from 'here' to 'there' is established.

All of these structures elaborate and depend on the potential for relating and qualifying inherent in the interaction of the one of 'I' and the zero of the point. Zero and one together make it possible to conduct experience and to reduce it. The range of possibilities is funneled through the available structures of experience; that is, the six senses.

The actual interplay of the senses with zero and one emerges in a special way. First comes the point. Zero, available within the point, has a distinctive feel. That 'feel' is preserved in the operation of each of the senses, from the seeing of sights and the hearing of sounds through the understanding of thoughts. In the moving from point to point inherent in each sense-field, the one who sees and hears emerges and is confirmed. The witness to experience takes form, unifying what is distinct and unique in each of the sensory modes. But this unity—the singularity of 'I'—depends on the distinctive feel of zero, symbol now of the wholeness of the sense field. Can we keep these two ways of understanding and experiencing in operation simultaneously? Can we engage the 'feel' of zero at the juncture of the 'I'?

One . . . Zero . . . Sixteen

'I' without zero leads to the sense of independent entities or things, their existence confirmed by the sense faculties that the 'I' possesses. As the I-centered 'from/to' operates in each sense-field, points take form in space and time. A structured and cohesive world emerges, suited to the knowledge available to us. Things are in the past, present, or future; things are empty or are objects; things have qualities and distinctions.

Things move; things arise, exist, and then pass away, so that they are no longer there.

At a deeper level, fundamental, 'I-centered' patterns cut across the distinctions of the sense-fields. The 'I' possesses and presents and provides. It gives advice to its own uniqueness. It confirms the singularity of one and the structure of 'here/there', of territory and identity.

What if we bring into play the symbol of zero? Now a certain openness manifests, allowing for what is new and creative. For instance, when the 'I' acknowledges the 'from/to' structure, a baseline is established. But zero shows us that the baseline is also a symbol, one that we are free to use in new ways. Suppose we decide to cross the baseline with another baseline. We have defined a midpoint, which means that we have gone beyond the original 'zero . . . one . . .two' and arrived at three. But actually, since we have two baselines, each crossing the other, we have twice three, or six.

Suppose (since zero allows for it) that these six points are connected to the six senses; that the sense-fields and the six points of the intersecting baselines connect with one another, sponsor one other, establish one other. In this way, we could say that one has become six. But one also remains one Thanks to the zero, one and six can be available together. When this 'together' comes into play, one and six give 16.

This way of arriving at 16 seems wholly arbitrary. Yet it also reflects a certain order. The zero unifies one and six, and this unity takes form in 16. We might imagine a progression and transition that leads to fifteen without ever bringing the zero into play. But with 16, there is a shift: Zero allows the point to be completely established. At one level, sixteen completes fifteen, and there, the move from one to six to sixteen makes no sense. At another level, however, 16 expresses the unique juncture of each being-point, and here 16 is the natural completion and integration of both zero and one. In 16, the distinctive feel of zero, as it manifests in each of the six sense-fields, is carried forward, not just as an arbitrary mark, but as the expression of an invariable and predictable order.

If 16 points toward such an order, we ought to be able to use it as an organizing principle. For instance, we could draw on 16 to chart out the field of interactions that human beings engage in. We cannot presently say what form this chart would take, but we would expect it to exhibit a fundamental order, like the periodic table of the elements, or the structure of DNA. In this structure, we would find the completion of zero and the foundation for fulfillment and accomplishment.

Sixteen Dimensions

Another way to arrive at the fullness of 16 would be to identify sixteen dimensions to reality. One approach is to start from the fourfold of space, time, knowledge, and being. Since each of these four in turn has four aspects (the space of space, the time of space, the knowledge of space, the being of space, and so on), we immediately arrive at 16. Invoking the power of this 16 as a symbol for the inner nature of all existence might allow for the possibility of finding the home for existence. From this foundation, it might be possible to trace the physical and mathematical laws that govern the unfolding of dimensionality in all possible domains.

How could we explore this prospect? One way would be to start once more with the six senses. If we saw each of the senses as a different dimension, we could quickly move on to countless open-ended junctures. For instance, we could speak in terms of 'from' and 'to' or investigate 'dependent on'. It is all at the level of symbols, so we have boundless freedom. The flavors of sensory awareness, each with its own tone or feel, is one direction to explore. But there are also the moods and directions of the senses, their derivations, reactions, and distinctions. There is the time of each sense, the space of each sense, the knowledge of each sense. As the movement of a twofold baseline unfolds in the space of zero, there are the aspects of dull or sharp, of clear or vague, of past, present, and future.

There is much to be done here—countless paths of inquiry. But we can say at the outset that tracing the shared qualities that the senses manifest into sixteen dimensions may allow us to bring the structures of zero, one, and 16 into the functional interplay of our daily experience. Then perhaps, we could find new and newly significant connections operating within the physical, logical, communicative, and symbolic realms.